# Spectrum Sensing for Cognitive Radio

# Spectrum Sensing for Cognitive Radio

## Fundamentals and Applications

### Kamal M. Captain
### Manjunath V. Joshi

CRC Press
Taylor & Francis Group
Boca Raton London New York

CRC Press is an imprint of the
Taylor & Francis Group, an **informa** business

MATLAB® is a trademark of The MathWorks, Inc. and is used with permission. The MathWorks does not warrant the accuracy of the text or exercises in this book. This book's use or discussion of MATLAB® software or related products does not constitute endorsement or sponsorship by The MathWorks of a particular pedagogical approach or particular use of the MATLAB® software.

First edition published 2022
by CRC Press
6000 Broken Sound Parkway NW, Suite 300, Boca Raton, FL 33487-2742

and by CRC Press
2 Park Square, Milton Park, Abingdon, Oxon, OX14 4RN

© 2022 Kamal M. Captain and Manjunath V. Joshi

CRC Press is an imprint of Taylor & Francis Group, LLC

Reasonable efforts have been made to publish reliable data and information, but the author and publisher cannot assume responsibility for the validity of all materials or the consequences of their use. The authors and publishers have attempted to trace the copyright holders of all material reproduced in this publication and apologize to copyright holders if permission to publish in this form has not been obtained. If any copyright material has not been acknowledged please write and let us know so we may rectify in any future reprint.

Except as permitted under U.S. Copyright Law, no part of this book may be reprinted, reproduced, transmitted, or utilized in any form by any electronic, mechanical, or other means, now known or hereafter invented, including photocopying, microfilming, and recording, or in any information storage or retrieval system, without written permission from the publishers.

For permission to photocopy or use material electronically from this work, access www.copyright.com or contact the Copyright Clearance Center, Inc. (CCC), 222 Rosewood Drive, Danvers, MA 01923, 978-750-8400. For works that are not available on CCC please contact mpkbookspermissions@tandf.co.uk

*Trademark notice:* Product or corporate names may be trademarks or registered trademarks and are used only for identification and explanation without intent to infringe.

ISBN: 978-0-367-54293-1 (hbk)
ISBN: 978-1-032-12605-0 (pbk)
ISBN: 978-1-003-08855-4 (ebk)

DOI: 10.1201/9781003088554

Typeset in Nimbus font
by KnowledgeWorks Global Ltd.

# Dedication

**Kamal M. Captain**
*To my mother, Mrs. Indu, and my father, Mr. Manhar*

**Manjunath V. Joshi**
*To Smita, Nidhi, Kishore and Ninad*

# Contents

Preface ............................................................................................................ xiii

Contributors ..................................................................................................... xv

List of Figures ................................................................................................ xvii

List of Tables .................................................................................................. xxi

Acronyms ...................................................................................................... xxiii

Acknowledgments ......................................................................................... xxv

**Chapter 1**     Fundamentals of Probability Theory ................................................ 1

         1.1    Introduction ............................................................................ 1
         1.2    Basics of Probability ............................................................... 2
                 1.2.1    Probability of an Event ............................................ 4
                          1.2.1.1    Axiomatic Definition ............................... 4
                          1.2.1.2    Relative Frequency Definition ................. 5
                          1.2.1.3    Classical Definition ................................. 5
                 1.2.2    Conditional Probability ........................................... 6
                 1.2.3    Independence of Events ........................................ 7
         1.3    Random Variable ................................................................... 7
                 1.3.1    Cumulative Distribution Function (CDF) ................ 8
                          1.3.1.1    Properties of Cumulative Distribution Function ................................................... 8
                 1.3.2    The Probability Density Function (PDF) ................. 9
                          1.3.2.1    Properties of Probability Density Functions ............................................................... 9
                 1.3.3    Joint Distribution and Density Function ................. 9
                 1.3.4    Conditional Probability Density Function ............. 10
                 1.3.5    Statistical Independence ....................................... 11
                 1.3.6    Moments of a Random Variable ............................ 11
                 1.3.7    Some Key Random Variables ............................... 14
                          1.3.7.1    Discrete Random Variables ................... 14
                          1.3.7.2    Continuous Random Variables ............... 15
                 1.3.8    The Markov and Chebyschev Inequalities .............. 22

vii

|  |  | 1.3.9 | The Sample Mean and the Laws of Large Numbers | 22 |
|---|---|---|---|---|
|  |  |  | 1.3.9.1 Weak Law of Large Numbers | 24 |
|  |  |  | 1.3.9.2 Strong Law of Large Numbers | 24 |
|  |  | 1.3.10 | Central Limit Theorem (CLT) | 24 |
|  | 1.4 | Stochastic Process | | 25 |
|  |  | 1.4.1 | Definition of Stochastic Process | 25 |
|  |  | 1.4.2 | Statistics of Stochastic Process | 26 |
|  |  | 1.4.3 | Stationarity | 28 |
|  |  |  | 1.4.3.1 Properties of Autocorrelation Function | 29 |
|  |  | 1.4.4 | Random Process through Linear System | 30 |
|  |  | 1.4.5 | Power Spectral Density (PSD) | 32 |
|  |  |  | 1.4.5.1 Properties of Power Spectral Density | 33 |
|  |  |  | 1.4.5.2 Output Spectral Density of an LTI System | 34 |
|  |  | 1.4.6 | Gaussian Random Process | 34 |
|  |  | 1.4.7 | White Noise | 35 |

**Chapter 2** Introduction ........................................................................ 37

    2.1 Cognitive Radio .............................................................. 38
    2.2 Spectrum Sensing ........................................................... 43
        2.2.1 Narrowband Spectrum Sensing ............................ 43
            2.2.1.1 Matched Filter Detection ..................... 44
            2.2.1.2 Cyclostationary Detection .................... 45
            2.2.1.3 Covariance-Based Detection ................ 46
            2.2.1.4 Eigenvalue-Based Detection ................. 46
            2.2.1.5 Energy Detection ................................. 46
        2.2.2 Wideband Spectrum Sensing ................................ 47
            2.2.2.1 Nyquist Wideband Spectrum Sensing ... 48
            2.2.2.2 Sub-Nyquist Wideband Spectrum Sensing ................................................. 48
        2.2.3 Cooperative Spectrum Sensing ............................. 49
        2.2.4 Machine-Learning-Based Spectrum Sensing ........ 51
    2.3 Book Contributions ........................................................ 52
    2.4 Tour of the Book ............................................................ 53

**Chapter 3** Literature Review .............................................................. 55

    3.1 Narrowband Spectrum Sensing ...................................... 55
    3.2 Wideband Spectrum Sensing .......................................... 58
    3.3 Cooperative Spectrum Sensing ....................................... 60
    3.4 Machine-Learning-Based Spectrum Sensing .................. 62

Contents ix

# PART I  Narrowband Spectrum Sensing

**Chapter 4**    Energy-Detection-Based Spectrum Sensing over Generalized Fading Model .................................................................. 67

     4.1   System and Channel Models ........................................... 68
         4.1.1   Energy Detection (ED) ........................................ 68
         4.1.2   $\eta$-$\lambda$-$\mu$ Fading Model ...................................... 70
     4.2   Average Probability of Detection over $\eta$-$\lambda$-$\mu$ Fading Channel ........................................................................... 70
         4.2.1   No Diversity ........................................................ 70
         4.2.2   Square Law Selection (SLS) Diversity .................. 72
         4.2.3   Cooperative Spectrum Sensing ............................ 74
     4.3   Average Probability of Detection over Channels with $\eta - \lambda - \mu$ Fading and Shadowing ........................................ 76
     4.4   Results and Discussion ................................................... 77
     4.5   Conclusion ....................................................................... 82

**Chapter 5**    Generalized Energy Detector in the Presence of Noise Uncertainty and Fading ................................................................. 83

     5.1   System Model .................................................................. 84
     5.2   Noise Uncertainty Model ................................................. 85
     5.3   SNR Wall for AWGN Channel ....................................... 86
         5.3.1   No Diversity ........................................................ 87
         5.3.2   pLC Diversity ..................................................... 89
         5.3.3   pLS Diversity ...................................................... 96
         5.3.4   CSS with Hard Combining ................................. 98
             5.3.4.1   OR Rule .............................................. 99
             5.3.4.2   AND Rule ......................................... 100
             5.3.4.3   $k$ Out of $M$ Combining Rule ................. 101
         5.3.5   CSS with Soft Combining .................................. 102
     5.4   SNR Wall for Fading Channel ....................................... 103
         5.4.1   No Diversity ...................................................... 103
         5.4.2   pLC Diversity ................................................... 105
         5.4.3   pLS Diversity .................................................... 106
         5.4.4   CSS with Hard Combining ............................... 106
             5.4.4.1   OR Combining .................................. 107
             5.4.4.2   AND Combining ............................... 107
         5.4.5   CSS with Soft Combining .................................. 108
     5.5   Results and Discussion ................................................. 108
         5.5.1   SNR Wall for AWGN Case ............................... 109
         5.5.2   SNR Wall for Fading Case ................................ 112

|  |  | 5.5.3 | Effect of Noise Uncertainty and Fading on Detection Performance ............................................. 114 |
|---|---|---|---|
|  |  | 5.5.4 | Effect of $p$ ............................................................. 116 |
|  | 5.6 | Conclusion ........................................................................ 117 |

# PART II  Wideband Spectrum Sensing

**Chapter 6**  Diversity for Wideband Spectrum Sensing under Fading ........... 121

        6.1  System Model and Performance Metrics ............................ 122
        6.2  Detection Algorithms ........................................................ 124
                6.2.1  Channel-by-Channel Square Law Combining (CC-SLC) ............................................................. 125
                6.2.2  Ranked Square Law Combining (R-SLC) Detection ............................................................... 126
                6.2.3  Ranked Square Law Selection (R-SLS) Detection ............................................................... 127
        6.3  Approximation of Decision Statistic .................................... 129
                6.3.1  PDF for SLC Diversity ......................................... 130
                        6.3.1.1  Without Using Approximation .............. 130
                        6.3.1.2  Using Approximation ............................ 130
                6.3.2  PDF for SLS Diversity .......................................... 132
                        6.3.2.1  Without Using Approximation .............. 132
                        6.3.2.2  Using Approximation ............................ 133
        6.4  Theoretical Analysis of Detection Algorithms .................... 133
                6.4.1  Channel-by-Channel Square Law Combining (CC-SLC) ............................................................. 133
                6.4.2  Theoretical Analysis for R-SLC ........................... 135
                6.4.3  Theoretical Analysis of R-SLS ............................. 138
        6.5  Results and Discussion ...................................................... 139
        6.6  Conclusion ........................................................................ 151

**Chapter 7**  Cooperative Wideband Spectrum Sensing .................................. 153

        7.1  System Model and Performance Metrics ............................ 154
        7.2  Proposed CWSS Algorithms .............................................. 155
                7.2.1  Proposed Algorithm Based on Hard Combining ... 155
                7.2.2  Proposed Algorithm Based on Soft Combining .... 157
        7.3  Approximation to pdf of Decision Statistic ........................ 159
        7.4  Theoretical Analysis of the Detection Algorithms ............. 162
                7.4.1  Theoretical Analysis for Algorithm 4 .................... 162
                        7.4.1.1  Performance Using Any Value of $M$ with Fixed $L$ ............................................ 164
                        7.4.1.2  Performance Using Any Value of $L$ with Fixed $M$ .......................................... 165

|  |  | 7.4.2 | Theoretical Analysis for Algorithm 5 ..................... 165 |
|---|---|---|---|
|  | 7.5 | Results and Discussion ......................................................... 168 |
|  |  | 7.5.1 | Experimentations Using Algorithm 4 .................... 168 |
|  |  | 7.5.2 | Experimentations Using Algorithm 5 .................... 174 |
|  | 7.6 | Conclusion ............................................................................ 178 |

**Chapter 8**   Conclusions and Future Research Directions .............................. 179

       8.1   Conclusions .......................................................................... 179
       8.2   Future Research Directions .................................................. 180

**Appendix A**   Appendix for Chapter 1 ............................................................ 183

       A.1   Proof for Markov Inequality ................................................ 183
       A.2   Proof Central Limit Theorem .............................................. 184
       A.3   Characteristic Function of Gaussian Random Variable ...... 185

**Appendix B**   Appendix for Chapter 4 ............................................................ 187

       B.1   Derivation for $P_F(\tau)$ in Eq. (4.3) ......................................... 187

**Appendix C**   Appendix for Chapter 5 ............................................................ 189

       C.1   Derivation for $\bar{P}_{D,plc}$ in Eq. (5.27) ........................................ 189
       C.2   Derivation for $\bar{P}_D^{Nak}$ in Eq. (5.77) ........................................ 192
       C.3   Derivation for $\bar{P}_{D,plc}^{Nak}$ in Eq. (5.81) ..................................... 193

**Appendix D**   Appendix for Chapter 6 ............................................................ 195

       D.1   Proof for Convergence of PDF of SLC under Nakagami
            Fading Channel in Eq. (6.12) ............................................... 195
       D.2   Derivation of PDF of SLS under Nakagami Fading in
            Eq. (6.17) .............................................................................. 195
       D.3   Proof for Convergence of PDF of SLS under Nakagami
            Fading in Eq. (6.17) ............................................................. 196
       D.4   Derivation of PDF in Eq. (6.13) ........................................... 197
       D.5   Derivation of Eq. (6.34) ....................................................... 198
       D.6   Derivation of PDF in Eq. (6.36) ........................................... 199
       D.7   Theoretical Analysis of R-SLC for $L=3$ ........................... 199

**Appendix E**   Some Special Functions ............................................................ 203

       E.1   Gamma Function .................................................................. 203
       E.2   Lower Incomplete Gamma Function ................................... 203
       E.3   Upper Incomplete Gamma Function ................................... 204
       E.4   Generalized Marcum Q-Function ........................................ 204
       E.5   Bessel Function of the First Kind ........................................ 204
       E.6   Modified Bessel Function of the First Kind ........................ 205

E.7 Confluent Hypergeometric Function ................................................. 205
E.8 Confluent Hypergeometric Function of the Second Kind .. 206
E.9 Unit Step Function ............................................................................ 206
E.10 Q-Function ........................................................................................ 206
E.11 Error Function .................................................................................. 207
E.12 Polylogarithm ................................................................................... 207

**Bibliography ........................................................................................................ 209**

**Index .................................................................................................................... 227**

# Preface

Rapid growth in new wireless communication services and applications has resulted in a continuous increase in the demand for radio frequency (RF) spectrum. Most of the available RF spectrum has already been licensed to the existing wireless systems. On the other hand, it is found that the spectrum is significantly underutilized due to the static frequency allocation to the dedicated users, and hence the spectrum holes or spectrum opportunities arise. Considering the scarce RF spectrum, supporting new services and applications is a challenging task that requires innovative technologies capable of providing new ways of exploiting the available radio spectrum. Cognitive radio (CR) has received immense research attention, both in academia and industry, as it is considered a promising solution to spectrum scarcity by introducing the notion of opportunistic spectrum usage. A CR is a device that senses the spectrum of licensed users (also known as primary users) for spectrum opportunities and transmits its data only when the spectrum is sensed to be not occupied. The process of identifying the empty licensed spectrum for opportunistic usage is termed in the literature as "Spectrum Sensing." For the efficient utilization of the spectrum, while limiting the interference to the licensed users, the CR should be able to sense the spectrum occupancy quickly and accurately. This makes spectrum sensing one of the main functionalities of the cognitive radio. Spectrum sensing is a hypothesis-testing problem, where the goal is to test whether the primary user is inactive (the null or noise-only hypothesis) or not (the alternate or signal-present hypothesis). Spectrum sensing can be broadly classified into two types, namely, narrowband and wideband sensing. Narrowband sensing is used for finding the occupancy status of a single licensed band, whereas wideband sensing deals with the scenario where multiple licensed bands are sensed for spectrum opportunities. This book discusses both types of spectrum sensing in depth.

## ABOUT THE BOOK

This book is primarily designed for senior undergraduate students, postgraduates, researchers, and industry personnel working in the areas of signal processing, next-generation wireless communication, and detection and estimation. The book will introduce readers to cognitive radio and the recent research in its main functionality, that is, spectrum sensing. The cognitive radio is the essential enabling technology for next-generation wireless communication like 5G and beyond. It guides the readers from the basic spectrum sensing algorithm for narrowband spectrum sensing to the more advanced wideband spectrum sensing, including cooperative sensing. In addition to that, the book discusses practical issues associated with implementing them in practice, like fading and noise uncertainty. Each chapter is equipped with rich mathematical proofs and derivations. We have also added appendices to help readers understand and derive the equations used in different chapters of the book.

## CONTENTS AND COVERAGE

Spectrum sensing is a hypothesis-testing problem where knowledge of probability theory is essential. In this book, we start by giving a brief introduction to probability theory in Chapter 1. This chapter is introduced to assist the reader with a ready reference to some of the basic probability, random variables, and random process concepts. The chapter starts with the basics of set theory on which the concept of probability is developed. The concept of probability is then extended for defining the random variables. We also discuss some of the key theorems related to random variables that are used in the subsequent chapters. Finally, the stochastic or random process is discussed in this chapter.

Chapter 2 introduces cognitive radio and its different elements. Spectrum sensing, which is one of CR's primary enabling technology, is defined, and different types of spectrum sensing techniques are briefly discussed. In Chapter 3, we discuss the literature survey on spectrum sensing.

Our primary focus is on analyzing the existing spectrum sensing algorithm considering practical scenarios and proposing novel spectrum sensing techniques in this book. Energy detection (ED), also known as conventional energy detection (CED) based spectrum sensing, is a very popular technique due to its simplicity and reduced computational complexity. In Chapter 4, we analyze the ED-based narrowband spectrum sensing over $\eta - \lambda - \mu$ fading channel model.

In Chapter 5, we discuss an interesting practical phenomenon called signal to noise ratio (SNR) wall for generalized energy detector (GED) considering no diversity, diversity, and cooperative narrowband sensing scenarios under noise uncertainty and fading. All the expressions derived are also validated using Monte Carlo simulations.

In the literature, antenna diversity to improve the detection performance of narrowband spectrum sensing is extensively studied. Chapter 6 proposes new detection algorithms that use square-law combining (SLC) and square-law selection (SLS) diversities for wideband spectrum sensing.

An alternative to antenna diversity is cooperative spectrum sensing, where multiple secondary users, also known as cooperating secondary users, collaborate by sharing their sensing information to detect the spectrum opportunities. In Chapter 7, we propose a novel detection algorithm for cooperative wideband spectrum sensing.

Finally, Chapter 8 discusses the conclusions and possible future research directions.

## KEYWORDS:

Cognitive radio, spectrum sensing, narrowband spectrum sensing, wideband spectrum sensing, energy detection, generalized energy detection, noise uncertainty, SNR wall, diversity, cooperative spectrum sensing, wideband spectrum sensing.

# Contributors

**KAMAL M. CAPTAIN:**

Dr. Kamal M. Captain received his Ph.D. in the area of Spectrum Sensing for Cognitive Radio from Dhirubhai Ambani Institute of Information and Communication Technology (DAIICT), Gandhinagar, India. Before this, he completed his M.Tech in Communication Systems from Sardar Vallabhbhai National Institute of Technology (SVNIT) and B.E degree in Electronics and Communication Engineering from Veer Narmad South Gujarat University (VNSGU), Surat, Gujarat, India. He is currently serving as an Assistant Professor at Sardar Vallabhbhai National Institute of Technology (SVNIT), Surat, Gujarat, India. Prior to joining SVNIT, he served as a senior engineer (signal processing) at eInfochips, Ahmedabad, India. He has been involved in active research in the areas of cognitive radio, wireless communication, signal processing, and machine learning and has several journals and international conference papers, including IEEE Transactions. He has also served as a reviewer for IEEE Transactions, letters, and top-tier conferences.

**MANJUNATH V. JOSHI:**

Prof. Manjunath V. Joshi received the B.E. degree from the University of Mysore, Mysore, India, and the M.Tech. and Ph.D. degrees from the Indian Institute of Technology Bombay (IIT Bombay), Mumbai, India. He is currently serving as a Professor and Dean of Research and Development with the Dhirubhai Ambani Institute of Information and Communication Technology, Gandhinagar, India. He has been involved in active research in the areas of Signal and Image Processing, Cognitive Radio, Computer Vision, and Machine Learning and has several publications in quality journals and conferences. He has co-authored four books entitled *Motion-Free Super Resolution* (Springer, New York-2005), *Digital Heritage Reconstruction Using Super-resolution and Inpainting* (Morgan and Claypool-2016), *Regularization in Hyperspectral Unmixing* (SPIE Press-2016), and the book entitled *Multi-resolution Image Fusion in Remote Sensing* (Cambridge University Press, UK-2019). Currently, he is contributing as a co-author of a book to be published by Springer in Remote Sensing, where seven positive reviews have been received. So far, nine Ph.D. students have graduated under his supervision. Dr. Joshi was a recipient of the Outstanding Researcher Award in Engineering Section by IIT Bombay in 2005 and the Dr. Vikram Sarabhai Award for 2006-2007 of information technology constituted by the Government of Gujarat, India. He served as a Program Co-Chair for the 3rd ACCV Workshop on E-Heritage, 2014, held in Singapore. He has also served as Visiting Professor at IIT Gandhinagar and IIIT Vadodara. He has visited Germany, Italy, France, Hong Kong, the USA, Canada, South Korea, and Indonesia and contributed to research in his area of expertise.

# List of Figures

1.1 Venn diagram representation of basic set operations. .................... 3
1.2 A function mapping $X(\zeta)$ from $S$ to real line. ........................ 8
1.3 Bernoulli random variable, (a) probability mass function, (b) cumulative distribution function. ........................................................ 14
1.4 Binomial random variable, (a) probability mass function, (b) cumulative distribution function. ........................................................ 15
1.5 Gaussian (Normal) random variable, (a) pdf, (b) cdf. .................... 16
1.6 Exponential random variable, (a) pdf, (b) cdf. ........................... 17
1.7 Chi-square random variable, (a) pdf, (b) cdf. ............................. 18
1.8 Non-central chi-square random variable, (a) pdf, (b) cdf. ............... 18
1.9 Rayleigh random variable, (a) pdf, (b) cdf. ................................ 19
1.10 Nakagami random variable, (a) pdf, (b) cdf. .............................. 20
1.11 Uniform random variable, (a) pdf, (b) cdf. ................................. 21
1.12 Gamma random variable, (a) pdf, (b) cdf. ................................. 22
1.13 A simple illustration of random process. .................................. 26
1.14 Autocorrelation function of slowly and rapidly fluctuating random process. ............................................................................ 30
1.15 Transmission of random process through linear time invariant system. ... 31
1.16 White noise, (a) power spectral density, (b) autocorrelation function. ... 36

2.1 Cognitive radio networks existing within a primary network. ........... 40
2.2 Dynamic Spectrum Management Framework for interweave network model [5, 131]. ................................................................... 41
2.3 Basic block diagram of narrowband sensing architecture at the secondary user. ................................................................................. 44
2.4 Basic block diagram of matched-filter-based spectrum sensing. ........ 45
2.5 Basic block diagram of energy-detection-based spectrum sensing. .... 46
2.6 Wideband spectrum consisting of $N$ frequency bands. .................. 47
2.7 Basic block diagram of Nyquist-sampling-based wideband spectrum sensing. ............................................................................. 48
2.8 Compressive-sensing-based sub-Nyquist wideband spectrum sensing using spectral domain energy detection approach. ........................ 49
2.9 Classification of cooperative spectrum sensing, (a) centralized, (b) distributed, and (c) relay-assisted. ............................................... 50
2.10 Block diagram of machine-learning-based spectrum sensing approach. ... 51

4.1 Block diagram of energy detection. .......................................... 68
4.2 Pictorial understanding of energy detection. The plot is obtained using $N = 10$ and $\gamma = 10\, dB$. .................................................... 69

| 4.3 | Block diagram of SLS diversity technique. | 73 |
|---|---|---|
| 4.4 | Centralized cooperative spectrum sensing. | 74 |
| 4.5 | Block diagram for CSS using OR combining. | 75 |
| 4.6 | $\bar{\gamma}$ vs $\bar{P}_D$ for no diversity considering for $P_F = 0.1$, $N = 4$, $\eta = 0.4$, $\lambda = 0.5$, $\mu = 1$. | 78 |
| 4.7 | $P_F$ vs $\bar{P}_M$ for no diversity case considering $\bar{\gamma} = 5\ dB$, $N = 4$, $\eta = 0.4$, $\lambda = 0.5$, $\mu = 1$. | 78 |
| 4.8 | $P_F$ vs $\bar{P}_M$ for other fading channels as a special case of $\eta$-$\lambda$-$\mu$ fading distribution with $\bar{\gamma} = 10\ dB$ and $N = 10$. | 79 |
| 4.9 | ROC plots under $\eta$-$\lambda$-$\mu$ fading channel using $\eta = 0.4$, $\lambda = 0.5$, $\mu = 1$, $N = 4$ for SLS diversity with $\bar{\gamma}_1 = 0\ dB$, $\bar{\gamma}_2 = 1\ dB$, $\bar{\gamma}_3 = 2\ dB$, $\bar{\gamma}_4 = 3\ dB$, and $\bar{\gamma}_5 = 4\ dB$ for different $P$ values. | 80 |
| 4.10 | ROC plots under $\eta$-$\lambda$-$\mu$ fading channel using $\eta = 0.4$, $\lambda = 0.5$, $\mu = 1$, $N = 4$ for cooperative spectrum sensing with $\bar{\gamma} = 0\ dB$ and $M$ cooperative secondary users. | 81 |
| 4.11 | $P_F$ vs $\bar{P}_D^{Shd}$ for composite multipath fading and shadowing channel for $k = 0.5$ and $k = 1$ with $\bar{\gamma} = 10\ dB$ and $N = 10$. | 82 |
| 5.1 | Block diagram of generalized energy detector. | 85 |
| 5.2 | Block diagram of pLC diversity technique. | 89 |
| 5.3 | $p$ VS. $\tilde{\gamma}_1$ for $\tilde{\gamma}_2 = 0.1$, $L_1 = L_2 = 1\ dB$. | 94 |
| 5.4 | Block diagram of pLS diversity technique. | 96 |
| 5.5 | Threshold ($\tau$) VS. detection probabilities for pLC diversity with two branches under AWGN channel using $L_1 = 1\ dB$, $L_2 = 1\ dB$, $\tilde{\gamma}_2 = 0.1$, $N = 10^7$ (a) for $p = 1$ and $\tilde{\gamma}_1 = 0.8914$ (b) for $p = 5$ and $\tilde{\gamma}_1 = 0.7153$. | 110 |
| 5.6 | Threshold ($\tau$) VS. detection probabilities for pLS diversity with two branches under AWGN channel using $L_1 = 0.5\ dB$, $L_2 = 0.3\ dB$, $\tilde{\gamma}_1 = 0.2$, $\tilde{\gamma}_2 = 0.1888$, $p = 2$, $N = 10^7$. | 110 |
| 5.7 | Plots of $\tau$ VS. $\bar{Q}_F$ and $\bar{Q}_D$ for $k$ out of $M$ combining rule. Here, $N = 10^6$, $M = 3$, $L_1 = 1\ dB$, $L_2 = 0.7\ dB$, $L_3 = 0.5\ dB$, $p = 2$, $\tilde{\gamma}_1 = 0.2$, $\tilde{\gamma}_2 = 0.3238$, and $\tilde{\gamma}_3 = 0.2836$. | 111 |
| 5.8 | Threshold ($\tau$) VS. detection probabilities for no diversity under Nakagami fading channel using $p = 2$, $N = 10^7$ for (a) no diversity with $L = 0.5\ dB$, $\bar{\gamma} = 1.82$, (b) pLC diversity using $L_1 = L_2 = 0.5\ dB$, $\tilde{\gamma}_1 = \tilde{\gamma}_2 = 0.67$, and (c) pLS diversity using $L_1 = 0.5\ dB$, $L_2 = 0.3$, $\tilde{\gamma}_1 = 0.1$, $\tilde{\gamma}_2 = 1.71$. | 113 |
| 5.9 | Threshold ($\tau$) VS. $\bar{Q}_F$ and $\bar{Q}_D$, for hard combining under fading channel using $N = 10^7$, $M = 3$, $p = 2$, (a) $k$ out of $M$ combining, (b) OR combining. | 114 |
| 5.10 | $\bar{P}_F$ Vs. $\bar{P}_D$ considering different scenarios, i.e., no fading and no noise uncertainty, with fading and no noise uncertainty, no fading with noise uncertainty ($L = 0.2\ dB$ and $L = 0.4$), and with fading and noise uncertainty ($L = 0.2\ dB$). Here, $N = 500$, $\bar{\gamma} = -10\ dB$, $p = 2$, $m = 2$. | 115 |

# List of Figures

5.11 $p$ Vs. $\bar{P}_D$ for no diversity, pLC diversity and pLS diversity for $\bar{P}_F = 0.1$ under fading and noise uncertainty considering (a) $L_1 = L_2 = L = 0.1\ dB$, $m = 2$, $\bar{\gamma}_1 = \bar{\gamma}_2 = \bar{\gamma} = -15\ dB$ with $N = 500$ and $N = 10^7$ and (b) $L_1 = L_2 = L = 0.5\ dB$, $m = 2$, $\bar{\gamma}_1 = \bar{\gamma}_2 = \bar{\gamma} = -5\ dB$ with $N = 100$. ........ 116

6.1 Block diagram of channel-by-channel square law combining (CC-SLC) detection. ........ 126
6.2 Block diagram of ranked square law combining (R-SLC) detection ........ 127
6.3 Block diagram of ranked square law selection (R-SLS) detection. ........ 128
6.4 Plots showing the actual and approximated pdf considering $N = 10$, $m_1 = 2$ for (a) $\bar{\gamma} = 0\ dB$, (b) $\bar{\gamma} = -5\ dB$, and (c) $\bar{\gamma} = -10\ dB$. ........ 131
6.5 $P_{EIO}$ Vs $P_{ISO}$ for CC-SLC considering $m_1 = 2$, $p = 0.1$, $L = 16$, $S_d = 1$, $I_d = 0$, $N = 10$, $\bar{\gamma} = -5\ dB$ for different number of diversity branches $P$. ........ 140
6.6 $P_{EIO}$ VS $P_{ISO}$ for different $P$ considering $m_1 = 2$, $p = 0.1$, $S_d = 1$, $I_d = 0$, $L_d = 1$, $N = 10$, $\bar{\gamma} = -5\ dB$ for R-SLC with $L = 16$. ........ 141
6.7 $P_{EIO}$ VS $P_{ISO}$ for different $P$ considering $m_1 = 2$, $p = 0.1$, $S_d = 1$, $I_d = 0$, $L_d = 1$, $N = 10$, $\bar{\gamma} = -5\ dB$ for R-SLS with $L = 2$. ........ 142
6.8 $P_{EIO}$ Vs $P_{ISO}$ showing the comparison of CC-SLC with R-SLC considering $m_1 = 2$, $p = 0.1$, $L = 8$, $S_d = 1$, $I_d = 0$, $L_d = 1$, $N = 10$, $\bar{\gamma} = -5\ dB$. 143
6.9 $P_{EIO}$ VS $P_{ISO}$ showing the comparison of R-SLC with R-SLS considering $m_1 = 2$, $p = 0.1$, $L = 2$, $P = 4$, $S_d = 1$, $I_d = 0$, $L_d = 1$, $N = 10$, $\bar{\gamma} = -5\ dB$. ........ 143
6.10 $P_{EIO}$ VS $P_{ISO}$ for varying $\bar{\gamma}$ using $m_1 = 2$, $N = 10$, $S_d = 1$, $I_d = 0$, $L_d = 1$, $P = 4$, $p = 0.1$, $L = 16$. ........ 144
6.11 $P_{EIO}$ VS $P_{ISO}$ for varying $p$ using $m_1 = 2$, $N = 10$, $S_d = 1$, $I_d = 0$, $L_d = 1$, $P = 4$, $\bar{\gamma} = -5\ dB$, $L = 16$. ........ 144
6.12 $P_{EIO}$ VS $P_{ISO}$ for varying $L$ using $m_1 = 2$, $N = 10$, $S_d = 1$, $I_d = 0$, $L_d = 1$, $P = 4$, $\bar{\gamma} = -5\ dB$, $p = 0.1$. ........ 146
6.13 $P_{EIO}$ VS $P_{ISO}$ plots considering different $p$ for channels in the partial band for $m_1 = 2$, $L = P = 8$, $S_d = 1$, $I_d = 0$, $L_d = 1$, $N = 10$, $\bar{\gamma} = -5\ dB$. ... 147
6.14 $P_{EIO}$ VS $P_{ISO}$ for R-SLC using theoretical analysis and Monte Carlo simulation considering $m_1 = 2$, $p = 0.1$, $S_d = 1$, $I_d = 0$, $L_d = 1$, $N = 10$, $L = P = 4$, and $\bar{\gamma} = -5\ dB$. ........ 148
6.15 $P_{EIO}$ VS $P_{ISO}$ for R-SLS using theoretical analysis and Monte Carlo simulation considering $m_1 = 2$, $p = 0.1$, $S_d = 1$, $I_d = 0$, $L_d = 1$, $N = 10$, $P = L = 2$, and $\bar{\gamma} = 0\ dB$. ........ 148
6.16 $P_{EIO}$ VS $P_{ISO}$ for varying $L_d$ using $m_1 = 2$, $N = 10$, $P = 4$, $\bar{\gamma} = 0\ dB$, $S_d = 1$, $I_d = 0$, $p = 0.1$, $L = 8$. ........ 149
6.17 $P_{EIO}$ VS $P_{ISO}$ for varying $S_d$ using $m_1 = 2$, $N = 10$, $P = 4$, $\bar{\gamma} = 0\ dB$, $L_d = 4$, $p = 0.1$, $L = 8$, $I_d = 0$. ........ 150
6.18 $P_{EIO}$ VS $P_{ISO}$ for varying $I_d$ using $m_1 = 2$, $N = 10$, $P = 4$, $\bar{\gamma} = 0\ dB$, $S_d = 4$, $p = 0.5$, $L = 16$, $L_d = 8$. ........ 150

7.1 Block diagram of proposed detection algorithm with $L = 2$, $M = 3$, $S_d = L_d = X_d = 1$, $I_d = 0$, $F_d = 1$. ........ 156

7.2  Block diagram for the proposed detection scheme with four CSUs ($M = 4$). ... 158
7.3  Plots showing the true and the approximated pdfs under $H_m = 1$ considering $N = 10$, $m_1 = 2$ for (a) $\bar{\gamma} = -5\ dB$ and (b) $\bar{\gamma} = 0\ dB$. ... 161
7.4  $P_{EIO}$ Vs. $P_{ISO}$ for the proposed algorithm using theoretical analysis and simulation considering $S_d = L_d = X_d = F_d = 1$, $I_d = 0$, $N = 10$, $m_1 = 2$, $\bar{\gamma} = 0\ dB$ (a) theoretical analysis in Section 7.4.1.1 with $M = 3$ and $L = 2$, (b) theoretical analysis in Section 7.4.1.2 with $L = 4$ and $M = 3$. ... 169
7.5  $P_{EIO}$ Vs. $P_{ISO}$ showing comparison of the proposed, RCD, R-SLC, and R-SLS considering $L = 2$, $M = 3$, $S_d = L_d = X_d = F_d = 1$, $I_d = 0$, $N = 10$, $m_1 = 2$, and $\bar{\gamma} = 0\ dB$. ... 170
7.6  $P_{EIO}$ Vs $P_{ISO}$ using $N = 10$, $S_d = L_d = X_d = F_d = 1$, $I_d = 0$, $p = 0.1$, $m_1 = 2$, $\gamma = 0\ dB$, $L = 2$ for different $M$. ... 172
7.7  $P_{EIO}$ Vs $P_{ISO}$ using $N = 10$, $S_d = L_d = X_d = F_d = 1$, $I_d = 0$, $p = 0.1$, $m_1 = 2$, $\gamma = 0\ dB$, $M = 3$ for different $L$. ... 172
7.8  $P_{EIO}$ Vs $P_{ISO}$ using $N = 10$, $S_d = 1$, $I_d = 0$, $p = 0.1$, $\gamma = 0\ dB$, $X_d = 2$, $L = 4$, $M = 8$, $L_d = 2$, $m_1 = 2$ for different $F_d$. ... 174
7.9  Plots for $P_{EIO}$ Vs. $P_{ISO}$ for the proposed algorithm using theoretical analysis and Monte Carlo simulation considering $M = 2$, $P = 2$, $L = 2$, $S_d = L_d = 1$, $I_d = 0$, $m_1 = 2$, and $\bar{\gamma} = -5\ dB$. ... 175
7.10 Plots of $P_{EIO}$ Vs. $P_{ISO}$ for the proposed, RCD [185], R-SLS [35], R-SLC [36], and Algorithm 4 considering $M = 2$, $P = 2$, $L = 2$, $S_d = L_d = 1$, $I_d = 0$, $m_1 = 2$, and $\bar{\gamma} = 0\ dB$. ... 175
7.11 Plots of $P_{EIO}$ Vs. $P_{ISO}$ for proposed algorithm considering $L = 4$, $P = 2$, $S_d = L_d = 1$, $I_d = 0$, $m_1 = 2$, and $\bar{\gamma} = -5\ dB$ for different number of CSUs $M$. ... 176
7.12 Plots of $P_{EIO}$ Vs. $P_{ISO}$ for proposed algorithm considering $M = 2$, $P = 2$, $S_d = L_d = 1$, $I_d = 0$, $m_1 = 2$, and $\bar{\gamma} = -5\ dB$ for different values of $L$. ... 177

# List of Tables

| | | |
|---|---|---|
| 2.1 | List of Few Unlicensed Frequency Bands in US [2] | 38 |
| 2.2 | Unlicensed Frequency Bands in UK [1] | 39 |
| 4.1 | Effect of Increasing Number of Diversity Branches on ($\bar{P}_D$) on the Performance | 79 |
| 4.2 | Effect of Increasing Number of Cooperating Secondary Users on ($\bar{P}_D$) on the Performance | 80 |
| 5.1 | Comparison of SNR Walls for Hard Combining. Here, $M = 3, L_1 = 1\ dB, L_2 = 0.7\ dB,$ and $L_3 = 0.5\ dB.$ | 102 |
| 5.2 | Comparison of SNR Walls for Hard Combining under Nakagami Fading. Here, $M = 3, L_1 = 1\ dB, L_2 = 0.7\ dB,$ and $L_3 = 0.5\ dB.$ | 108 |
| 6.1 | Table of Notations | 123 |
| 6.2 | Effect of Increasing Number of Diversity Branches $P$ on $P_{ISO}$ for CC-SLC | 140 |
| 6.3 | Effect of Increasing Number of Diversity Branches $P$ on $P_{ISO}$ for R-SLC | 141 |
| 6.4 | Effect of Increasing Number of Diversity Branches $P$ on $P_{ISO}$ for R-SLS | 141 |
| 6.5 | Effect of Increasing Average SNR $\bar{\gamma}$ on $P_{ISO}$ for R-SLC | 145 |
| 6.6 | Effect of Increasing Occupancy Probability $p$ on $P_{ISO}$ for R-SLC | 145 |
| 6.7 | Effect of Increasing Number of Channels in the Partial Band $L$ on $P_{ISO}$ for R-SLC | 145 |
| 6.8 | Occupancy Probabilities for $L = 8$ Subbands | 146 |
| 6.9 | Effect of Different Occupancy Probabilities on $P_{ISO}$ for R-SLC | 147 |
| 6.10 | Effect of Increasing $L_d$ on $P_{ISO}$ for R-SLC | 149 |
| 6.11 | Effect of Different $S_d$ on $P_{ISO}$ for R-SLC | 151 |
| 6.12 | Effect of Different $I_d$ on $P_{ISO}$ for R-SLC | 151 |
| 7.1 | Comparison of Algorithm 4 with Other Algorithms | 171 |
| 7.2 | Effect of Increasing $M$ on Algorithm 4 | 171 |
| 7.3 | Effect of Increasing $L$ on Algorithm 4 | 173 |
| 7.4 | Effect of Increasing $F_d$ on Algorithm 4 | 173 |
| 7.5 | Comparison of Algorithm 5 with Other Algorithms | 176 |
| 7.6 | Effect on Increasing $M$ | 177 |
| 7.7 | Effect on Increasing $L$ | 177 |

# Acronyms

| | |
|---|---|
| ADC | Analog-to-digital converter |
| AIC | Analog-to-information converter |
| ASSS | Adaptive spectrum sensing strategy |
| AUC | Area under the receiver operating characteristic |
| AWGN | Additive white Gaussian noise |
| BPF | Bandpass filter |
| CAF | Cyclic autocorrelation function |
| CCD | Channel-by-channel detection |
| CCDF | Complementary cumulative distribution function |
| CC-SLC | Channel-by-channel square law combining |
| CDF | Cumulative distribution function |
| CED | Conventional energy detection |
| CLT | Central limit theorem |
| CLDNN | Convolutional long short-term deep neural network |
| CNN | Convolution neural network |
| CR | Cognitive radio |
| CSI | Channel state information |
| CSU | Cooperating secondary user |
| DCS | Deep cooperative sensing |
| DFT | Discrete Fourier transform |
| DOF | Degree of freedom |
| DRL | Deep reinforcement learning |
| DSMF | Dynamic spectrum management framework |
| DVB-T | Digital video broadcasting-terrestrial |
| ED | Energy detection |
| EGC | Equal gain combining |
| EM | Expectation maximization |
| ER | Effective rate |
| FC | Fusion center |
| FCC | Federal communications commission |
| FFT | Fast Fourier transform |
| FHT | Fuzzy hypotheses testing |
| GED | Generalized energy detection |
| GLRT | Generalized likelihood ratio test |
| GMM | Gaussian mixture model |
| HMM | Hidden Markov model |
| IID | Independent and identically distributed |
| ITU | International telecommunication union |
| JDE | Joint detection and estimation |

| | |
|---|---|
| **JPDF** | Joint probability density function |
| **KNN** | K-nearest-neighbor |
| **LTI** | Linear time invariant |
| **LNA** | Low noise amplifier |
| **MAC** | Media access control |
| **MC** | Monte Carlo |
| **MF** | Matched filter |
| **MGF** | Moment generating function |
| **MPTP** | Multiple primary transmit power |
| **MRC** | Maximal ratio combining |
| **MSE** | Mean-square estimation error |
| **MWC** | Modulated wideband converter |
| **NBC** | Naive Bayes classifier |
| **NU** | Noise uncertainty |
| **OFDM** | Orthogonal frequency division multiplexing |
| **PBNS** | Partial band Nyquist sampling |
| **PDF** | Probability density function |
| **PMF** | Probability mass function |
| **PSD** | Power spectral density |
| **PU** | Primary user |
| **QoS** | Quality of service |
| **RCD** | Ranked channel detection |
| **RF** | Radio frequency |
| **ROC** | Receiver operating characteristic |
| **RSSS** | Random spectrum sensing strategy |
| **R-SLC** | Ranked square law combining |
| **R-SLS** | Ranked square law selection |
| **SBL** | Sparse Bayesian learning |
| **SCF** | Spectrum correlation function |
| **SLC** | Square law combining |
| **SLS** | Square law selection |
| **SNR** | Signal to noise ratio |
| **SPTF** | Spectrum policy task force |
| **SSC** | Spectrum sensing capability |
| **SSS** | Strict sense stationary |
| **SU** | Secondary user |
| **SVM** | Support vector machine |
| **UHF** | Ultra high frequency |
| **WSS** | Wideband spectrum sensing |
| **WGN** | White Gaussian noise |

# Acknowledgments

This book is the result of the encouragement and support of many wonderful people. It gives us great pleasure in expressing our gratitude to all those who have supported us and contributed in making it possible. We were immensely benefited from the comments and suggestions from the following people: Prof. K. S. Dasgupta, Prof. Deepak Ghodgaonkar, Prof. Hemant Patil, Dr. Yash Vasawada, Dr. Aditya Tatu, Dr. Laxminarayana Pillutla, and Prof. Suman Mitra. We thank the reviewers for their constructive suggestions. Thanks to Gauravjeet Singh from CRC Press for accepting the proposal. We are also thankful to Lakshay Gaba and his team members from CRC Press for the cooperation and strong support during the review and proofreading stages. We personally want to thank all those who have helped us throughout the journey.

**Kamal M. Captain:**
Although the work described in this book was performed independently, I never would have been able to complete it without the support of many wonderful people. I would like to offer my sincere thanks to them.

I would like to begin by expressing my sincere gratitude to my Ph.D. supervisor, Prof. Manjunath V. Joshi, with whom I have learned immensely and who has had a strong influence in my development as a researcher and a teacher. He has constantly encouraged me to ensure that I remain focused on achieving my goal. I am grateful to him for patiently supervising and directing my work, fruitful discussions, providing learning opportunities on a number of occasions, and helping me throughout all the different steps of my doctoral research endeavor for the past few years. Prof. Joshi's achievements, his work ethics, and his keen eye for every important detail have been an inspiration throughout all the years I have worked with him.

On a broader note, I wish to acknowledge all the professors of DA-IICT who have inspired me directly or indirectly. I would like to thank all my colleagues at SVNIT, Surat for their constant support and encouragement.

I made a lot of new friends at DA-IICT who helped me in many steps of my study. I thank all of them for everything that they did for me. I am thankful to Krishna Gopal, Rishikant, and Jignesh for helping me in deriving lengthy mathematical expressions. Thanks to Pramod, Hardik, Milind, Idrish, Tanvina, Zaki, Avni, my roommates, Sumukh, Parth, Nirmesh, and Gaurav, and other Ph.D. batchmates at DA-IICT Prashant, Nupur, and Vandana for their kind cooperation. I want to thank my tuition teacher, Mr. Fareed, who planted the seed of my love for mathematics, and my physics teacher, Mr. Jayesh Tandel, for helping me develop positive thinking. I am also thankful to all my friends from SVNIT, MGITER, and BM&BFWHS for their constant support and encouragement.

I am deeply thankful to my mother, Indu, and my father, Manhar, for their love and support. Words cannot express how grateful I am to my family for all of the sacrifices they have made on my behalf. Your prayer for me was what sustained me thus far. I warmly thank my brother, Rohan, my sister-in-law, Khyati, and my in-laws, Anoj and Kumud, for their understanding and support in many aspects of my life. I personally want to thank my brother, Rohan, for always being there for me when I needed. I want to thank my niece, Yashvi (4 years old), for allowing me to use my laptop even though she has a lot of work to do on my laptop. I acknowledge my entire family for providing me with a healthy, educated atmosphere in our family.

This last word of acknowledgment I have saved for my dear wife, Shreya. I always fall short of words and felt impossible to describe her support in words. Without your unconditional support and encouragement, I could not have finished this book. I see myself unable to even express my feelings about the love and patience that I observed from you. Thanks a lot for bearing my busy schedule and not complaining. Finally, to the special one who is joining us soon. Thank you for motivating me to complete this book as early as possible so that we can spend more time together.

**Manjunath V. Joshi:**
I would like to thank Dr. Milind Padalkar, Dr. Rakesh Patel, Dr. Shrishail Gajbhar, Dr. Sonam Nahar, and Jignesh for their kind help during the writing of this book. I wish to express my deep sense of gratitude to my family members, Smita, Nidhi, Kishore, and Ninad, my sisters and brothers, for their constant love and encouragement during the preparation of this book. I am grateful to my academic mentors and Professors K. V. V. Murthy, S. Chaudhuri, and P. G. Poonanacha for their inspiration and support.

Finally, we wish to thank all who helped us directly or indirectly for their support during the entire process of publishing this book.

# 1 Fundamentals of Probability Theory

In this chapter, the fundamentals of probability, random variables, and random process are discussed briefly. This chapter is introduced to give a quick overview of these topics to the readers[1].

## 1.1 INTRODUCTION

A signal can be classified as being deterministic or random. For the deterministic signal, there is no uncertainty associated with its value at a particular time instant. A mathematical formula exactly defines the deterministic signals. Random signals are unpredictable; that is, the future values of the random signals cannot be predicted with certainty even if the entire history of the signal is known.

Consider the signal $y(t) = A\sin(2\pi ft + \theta)$. If we assume the values of $A$, $f$, and $\theta$ to be constant and known, we can determine the value of $y(t)$ for all values of $t$. The values of these parameters, if not known, can be obtained by observing the signal for a short period of time. Now, if we assume that the signal $y(t)$ be the output of a signal generator having inferior frequency stability and if this signal generator is set to produce the sinusoid of frequency $f$, the actual frequency output will be $f' \in (f + \Delta f)$. This generated signal may not remain same and could vary over time. It may not be of great use to observe the signal over a long period of time to predict future values. We can say that the output of signal generator varies randomly. Another example of a random signal is the received signal at the receiving antenna in the wireless communication scenario. The received signal amplitude at the receiving antenna varies unpredictably because of the wireless channel effects such as path loss, fading, shadowing, and Doppler shift. These wireless channel impairments are random and not in our control.

Few signals experienced in the real world exhibit statistical regularity even though their exact behavior remains unpredictable. For example, once again consider the wireless signal amplitude received at the receiving antenna after propagation through the channel. It is difficult to predict the exact value of the received signal amplitude at the output of the receiving antenna at a particular time, but we can certainly observe that the average signal amplitudes over successive time intervals do not vary significantly. Let us consider the simple example of throwing a fair die. There are six possible outcomes, but we do not know in advance what will be the outcome of

---

[1]The discussion given here is minimal and focused on giving quick reference to the reader. Readers are encouraged to go through books dedicated to probability theory to get detailed discussion on these topics [102, 112, 140, 150, 156, 157].

DOI: 10.1201/9781003088554-1

when you throw a die. Our experience tells us that if we throw a die large number of times and note the outcomes, any particular outcome will be observed for one-sixth of the time. If this does not happen, then we suspect the fairness of the die.

Statistical regularity in some random physical signals can be verified experimentally. Hence, it is necessary to have a mathematical tool to analyze and characterize such random signals mathematically. To analyze the random signals, we have to understand random variables. The mathematical topics that are key to understand the random nature of the signal are probability theory, random variables, and random processes. This chapter discusses these topics and some fundamental theorems that are useful for understanding other chapters.

## 1.2 BASICS OF PROBABILITY

The following set of terminology related to random experiment is used in the probability theory. The result of an experiment is called an **outcome**. For example, in the experiment of tossing a coin, the outcome would be head or tail. The **random experiments** are those whose outcomes are not known in advance. For example, drawing a card from a deck, throwing a die, measuring a noise voltage, etc. An outcome or set of outcomes of a random experiment is called a **random event**. For example, in the experiment of drawing a card, an event could be the 'drawn card is an ace' or 'drawn card is of heart.' The **sample space** of an event is the set of all the possible outcomes of a random experiment. The sample space is denoted by $S$ or $\Omega$, and is also called the **certain event**. The elements of certain events are called experimental outcomes, and the events are a subset of sample space. The **impossible event** or **null event** is the one which contains no outcomes and hence never occurs. The null event is denoted by $\Phi$.

We can make use of **set operations** to combine events to obtain other events, or we can express complex events using a combination of simple events. The union of two events $A$ and $B$ is the set of all outcomes that are either in $A$ or in $B$, or in both $A$ and $B$. The union of $A$ and $B$ is denoted by $A \cup B$. We say event $A \cup B$ has occurred if either $A$, or $B$, or both $A$ and $B$ have occurred. The **intersection** of events $A$ and $B$ is defined as the set of outcomes that are in both $A$ and $B$, and is denoted by $A \cap B$. The event $A \cap B$ is said to have occurred if both $A$ and $B$ have occurred. If the intersection of events $A$ and $B$ is null event, i.e., $A \cap B = \Phi$, then these two events are called **mutually exclusive**. Mutually exclusive events cannot occur simultaneously. The **partitions** $U$ of a set $S$ is a collection of mutually exclusive subsets $A_i$ of $S$ whose union equals $S$. The **complement** of an event $A$ consists of all the outcomes that are not members of event $A$ and it is denoted by $A^c$ or $\bar{A}$. The event $A$ and $\bar{A}$ cannot occur simultaneously. If an event $A$ is a subset of event $B$, i.e., $A \subset B$, then event $B$ will occur whenever event $A$ occurs. We also say that event $A$ implies an event $B$. If events $A$ and $B$ contain the same outcomes, then we say events $A$ and $B$ are equal, i.e., $A = B$.

Fig. 1.1 shows few basic set operations on the sets $A$ and $B$ using the Venn diagram. In this diagram, a rectangle is used to represent the sample space $S$, the circle is used to represent the event (a subset of $S$), and the shaded region represents the various events.

# Fundamentals of Probability Theory

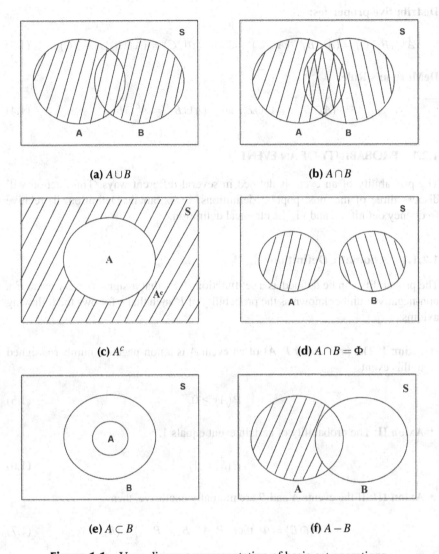

**Figure 1.1** Venn diagram representation of basic set operations.

Some useful properties of set operations are described below:

**Commutative properties:**

$$A \cup B = B \cup A \text{ and } A \cap B = B \cap A. \tag{1.1}$$

**Associative properties:**

$$A \cup (B \cup C) = (A \cup B) \cup C \text{ and } A \cap (B \cap C) = (A \cap B) \cap C. \tag{1.2}$$

**Distributive properties:**

$$A \cup (B \cap C) = (A \cup B) \cap (A \cup C) \text{ and } A \cap (B \cup C) = (A \cap B) \cup (A \cap C). \quad (1.3)$$

**DeMorgan's Rule:**

$$(A \cap B)^c = A^c \cup B^c \text{ and } (A \cup B)^c = A^c \cap B^c. \quad (1.4)$$

### 1.2.1 PROBABILITY OF AN EVENT

The probability of an event is defined in several different ways. This section will discuss three of the most popular definitions: i) Axiomatic Definition, ii) relative frequency definition, and iii) the classical definition.

#### 1.2.1.1 Axiomatic Definition

The probability can be defined as a set function $P(E)$ that assigns to every event $E$ a non-negative number known as the probability of $E$ such that it follows the following axioms:

- **Axion I**: The probability $P(A)$ of an event $A$ is a non-negative number assigned to this event

$$P(A) \geq 0. \quad (1.5)$$

- **Axion II**: The probability of certain event equals 1.

$$P(S) = 1. \quad (1.6)$$

- **Axion III**: If the events $A$ and $B$ are mutually exclusive, then

$$P(A \cap B) = \Phi \text{ then } P(A \cup B) = P(A) + (B). \quad (1.7)$$

- **Axion III'**: If $A_1, A_2, \ldots$ is a sequence of events such that $A_i \cap A_j = \Phi$ for all $i \neq j$, then

$$P\left[\bigcup_{k=1}^{\infty} \right] = \sum_{k=1}^{\infty} P(A_k). \quad (1.8)$$

To deal with finite sample space experiments, only the first three axioms I, II, and III are sufficient. To deal with experiments with infinite sample space, we need to replace Axiom III with Axiom III'. Note that Axiom III is a special case of Axiom III' with $A_k = \Phi$ for $k \geq 3$. Thus, in reality, we need only Axioms I, II, III'.

# Fundamentals of Probability Theory

## 1.2.1.2 Relative Frequency Definition

Let a random experiment be conducted $n$ times. If the event $A$ occurs $n_A$ times, then as per relative frequency definition, the probability of an event $A$ is defined as the limit

$$P(A) = \lim_{n \to \infty} \frac{n_A}{n}. \tag{1.9}$$

Here, the term $\left(\frac{n_A}{n}\right)$ represents the fraction of occurrence of $A$ in $n$ trials. When the value of $n$ is small, it is possible that the term $\left(\frac{n_A}{n}\right)$ will fluctuate badly. But as $n$ grows larger and larger, we expect that $\left(\frac{n_A}{n}\right)$ converges to a definite value. For example, let us consider the experiment of tossing of a coin and let an event $A$ be "the outcome of toss is head." If $n$ is small, let say 100, then $\left(\frac{n_A}{n}\right)$ may not deviate from $\frac{1}{2}$ by more than say 10%. But as $n$ grows larger and larger, the term $\left(\frac{n_A}{n}\right)$ converges towards $\frac{1}{2}$. The probability defined using relative frequency interpretation must follow all the axioms given in Section 1.2.1.1.

## 1.2.1.3 Classical Definition

According to the classical definition, the probability $P(A)$ of the event $A$ is found without any experimentation. The probability $P(A)$ is defined as the ratio

$$P(A) = \frac{N_A}{N}, \tag{1.10}$$

where, $N$ represents the number of possible outcomes and $N_A$ is the number of favorable outcomes to event $A$. Here, it is assumed that all the outcomes are equally likely, i.e., they have the same chance to occur. For example, if we consider the experiment of rolling a fair die and an event $A$ as the even outcome, the total number of possible outcomes is six ($N = 6$), and the favorable outcomes to event $A$ are three ($N_A = 3$). Hence, the probability $P(A)$ of an event $A$ is $\frac{3}{6} = \frac{1}{2}$. Once again, the probability must follow the three axioms defined in Section 1.2.1.1.

Using the axioms defined in Section 1.2.1.1, it is possible to derive some additional relationships:

- If $A \cap B \neq \Phi$, then $P(A \cup B) = P(A) + P(B) - P(A \cap B)$.
- Let $A_1, A_2, \ldots, A_n$ be random events such that

$$A_i \cap A_j = \Phi \text{ for } i \neq j \text{ and } A_1 + A_2 + \cdots + A_n = S. \tag{1.11}$$

then,

$$P(A) = P(A \cap A_1) + P(A \cap A_2) + \cdots + P(A \cap A_n), \tag{1.12}$$

where, $A$ is any event on the sample space and $A_1, A_2, \ldots, A_n$ are said to be mutually exclusive and exhaustive.
- $P(\bar{A}) = 1 - P(A)$.

## 1.2.2 CONDITIONAL PROBABILITY

Conditional probability is a useful concept in probability theory. It is denoted by $P(B|A)$ and represents a probability of event $B$ occurring given that an event $A$ has occurred. In the real world, it is very much likely that the occurrence of an event $B$ is greatly affected by the occurrence of another event $A$. For example, a box contains three blue marbles and two red marbles. The event "second picked marble is blue" depends on what is picked in the first attempt. The conditional probability is defined in terms of joint probability $P(A \cap B)$ and the probability of the event $A$, and is given by

$$P(B|A) = \frac{P(A \cap B)}{P(A)}, \quad P(A) \neq 0, \text{ or} \qquad (1.13)$$

$$P(A \cap B) = P(B|A)P(A). \qquad (1.14)$$

If we interchange the role of $A$ and $B$, we have

$$P(A|B) = \frac{P(A \cap B)}{P(B)}, \quad P(B) \neq 0. \qquad (1.15)$$

Using Eq. (1.13) and Eq. (1.15), we can write

$$P(A \cap B) = P(B|A)P(A) = P(A|B)P(B). \qquad (1.16)$$

Using Eq. (1.16), we can write

$$P(B|A) = \frac{P(B)P(A|B)}{P(A)}, \quad P(A) \neq 0, \qquad (1.17)$$

Similarly,

$$P(A|B) = \frac{P(A)P(B|A)}{P(B)}, \quad P(B) \neq 0, \qquad (1.18)$$

Eq. (1.17) and Eq. (1.18) represent one form of Bayes' theorem or Bayes' rule.

Let $A_1, A_2, \ldots, A_n$ be a partition of a sample space $S$ and the event $B$ is another event on the same sample space. It can be shown that

$$P(A_j|B) = \frac{P(B|A_j)P(A_j)}{\sum_{i=1}^{n} P(B|A_j)P(A_j)}. \qquad (1.19)$$

Eq. (1.19) represents another form of Bayes' theorem.

# Fundamentals of Probability Theory

## 1.2.3 INDEPENDENCE OF EVENTS

Consider two events $A$ and $B$ on the sample space $S$. If the knowledge of the occurrence of event $B$ does not affect the probability of event $A$, then we say that the event $A$ and $B$ are independent of each other. We can then write

$$P(A|B) = P(A) \quad \text{or} \quad P(B|A) = P(B). \tag{1.20}$$

Using Eq. (1.13), we can write

$$P(B|A) = P(B) = \frac{P(A \cap B)}{P(A)}. \tag{1.21}$$

The problem with Eq. (1.21) is that the equation is not defined for $P(A) = 0$. Using Eq. (1.21), we will define the two events $A$ and $B$ as independent if

$$P(A \cap B) = P(A)P(B). \tag{1.22}$$

The definition of statistical independence can be generalized to more than two events. Suppose we have set of $n$ events denoted by $A_1, A_2, \ldots, A_n$. The events are said to be independent if and only if (iif) the probability of every intersection of $n$ or fewer events is equal to the product of its constituent probabilities. Thus the three events $A_1$, $A_2$, and $A_3$ are independent when

$$P(A_1 \cap A_2) = P(A_1)P(A_2), \tag{1.23}$$

$$P(A_2 \cap A_3) = P(A_2)P(A_3), \tag{1.24}$$

$$P(A_1 \cap A_3) = P(A_1)P(A_3), \quad \text{and} \tag{1.25}$$

$$P(A_1 \cap A_2 \cap A_3) = P(A_1)P(A_2)P(A_3). \tag{1.26}$$

## 1.3 RANDOM VARIABLE

Let us consider an experiment specified by a sample space $S$, the subset of $S$ called events, and the probability assigned to these events. We assign a real number $X(\zeta)$ to every outcome $\zeta$ of an experiment. Thus, we have created a transformation or a function $X$ whose domain is the sample space $S$ of a random experiment and whose range is a set of real numbers. To each experimental outcome $\zeta_i \in S$, the function $X$ assigns a real number $X(\zeta_i)$ as shown in Fig. 1.2. This function is called the random variable.

All random variables will be written in capital letters in this book. The symbol $X(\zeta)$ will indicate the number assigned to the outcome $\zeta$ and $X$ represents the rule of correspondence between elements of sample space $S$ and the real number assigned to it. In rest of the book, random variables are represented using boldface letters and the function of $\zeta$ is not explicitly mentioned for simplicity of notations. For example, $X$ is the table pairing two faces of a coin with two numbers 1 and $-1$. The domain of this function is the set $S = \{H, T\}$ and its range is $\{1, -1\}$. The expression $X(H)$ is the number 1. A random variable can be continuous, discrete, or mixed.

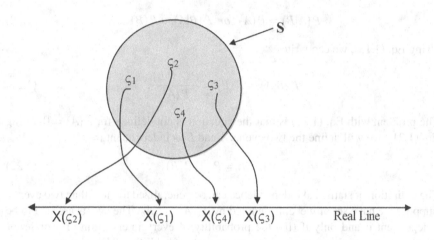

**Figure 1.2** A function mapping $X(\zeta)$ from $S$ to real line.

## 1.3.1 CUMULATIVE DISTRIBUTION FUNCTION (CDF)

Many a times we need to compute the $P\{X = x\}$, $P\{X \leq x\}$ or $P\{x_1 \leq X \leq x_2\}$. To compute these probabilities, we need to consider the cumulative distribution function (cdf). The number of elements of the set $S$ that are contained in the event $\{X \leq x\}$ changes as the number $x$ takes various values. The probability $P\{X \leq x\}$ of the event $\{X \leq x\}$ depends on the number $x$. This probability is denoted by $F_X(x)$ and is called the cumulative distribution function of random variable $X$ which is defined as

$$F_X(X) = P\{X \leq x\}, \text{ for } -\infty < x < \infty. \tag{1.27}$$

In Eq. (1.27), the variable $x$ can be replaced by any letter. For example, $F_X(y) = P\{X \leq y\}$.

### 1.3.1.1 Properties of Cumulative Distribution Function

The cumulative distribution function has the following properties:

1. $F_X(+\infty) = 1$ $F_X(-\infty) = 0$.
2. It is a nondecreasing function of $x$, i.e., if $x_1 < x_2$ then $F_X(x_1) \leq F_X(x_2)$.
3. If $F_X(x') = 0$ then $F_X(x) = 0$ for every $x \leq x'$.
4. $P\{X > x\} = 1 - F_X(x)$.

# Fundamentals of Probability Theory

5. The function $F_X(x)$ is continuous from right which gives

$$F_X(x^+) = \lim_{\varepsilon \to 0} F_X(x+\varepsilon) = F_X(x), \varepsilon > 0. \qquad (1.28)$$

6. $P\{x_1 < X \leq x_2\} = F_X(x_2) - F_X(x_1)$.
7. $P\{X = x\} = F_X(x) - F_X(x^-)$ and $P\{X = x\} = 0$ if random variable is continuous.
8. $P\{x_1 \leq X \leq x_2\} = F_X(x_2) - F_X(x_1^-)$ and for continuous random variable $F_X(x_1^-) = F_X(x_1)$.

The proofs of these properties are not discussed here because this chapter aims to provide the reader with a quick reference to the probability theory to better understand chapters to be discussed next.

## 1.3.2 THE PROBABILITY DENSITY FUNCTION (PDF)

The derivative of cumulative distribution function $F_X(x)$ is called the probability density function (pdf) of random variable $X$ and is denoted by $f_X(x)$. Thus we have

$$f_X(x) = \frac{dF_X(x)}{dx}. \qquad (1.29)$$

### 1.3.2.1 Properties of Probability Density Functions

Following are some fundamental properties of the probability density functions:

1. $f_X(x) \geq 0$.
2. $\int_{-\infty}^{+\infty} f_X(x)\,dx = 1$.
3. $P\{x_1 < X \leq x_2\} = \int_{x_1}^{x_2} f_X(x)\,dx$.

## 1.3.3 JOINT DISTRIBUTION AND DENSITY FUNCTION

Let us consider two random variables $X$ and $Y$. They can be characterized by the two-dimensional cdf given by

$$F_{X,Y}(x,y) = P\{X \leq x, Y \leq y\}. \qquad (1.30)$$

The two-dimensional cdf possess the following properties:

1. $F_{X,Y}(x,y) \geq 0$, $-\infty < x < \infty$, $-\infty < y < \infty$.
2. $F_{X,Y}(-\infty,y) = F_{X,Y}(x,-\infty) = 0$, $F_{X,Y}(+\infty,+\infty) = 1$.
3. $F_{X,Y}(\infty,y) = F_Y(y)$ and $F_{X,Y}(x,\infty) = F_X(x)$.
4. If $x_2 > x_1$ and $y_2 > y_1$, then $F_{X,Y}(x_2,y_2) \geq F_{X,Y}(x_2,y_1) \geq F_{X,Y}(x_1,y_2)$.

5. $P\{x_1 < X \le x_2, Y \le y\} = F_{X,Y}(x_2,y) - F_{X,Y}(x_1,y)$ and $P\{X \le x, y_1 < Y \le y_2\} = F_{X,Y}(x,y_2) - F_{X,Y}(x,y_1)$.
6. $P\{x_1 < X \le x_2, y_1 < Y \le y_2\} = F_{X,Y}(x_2,y_2) - F_{X,Y}(x_1,y_2) - F_{X,Y}(x_2,y_1) + F_{X,Y}(x_1,y_1)$.

The two-dimensional probability density function is given as

$$f_{X,Y}(x,y) = \frac{\partial^2}{\partial x \partial y} F_{X,Y}(x,y), \tag{1.31}$$

or

$$F_{X,Y}(x,y) = \int_{-\infty}^{y} \int_{-\infty}^{x} f_{X,Y}(z,w) \, dz \, dw. \tag{1.32}$$

The notion of joint cdf and pdf can be extended to the case of $n$ number of random variables where $n \ge 3$.

Given the joint pdf of $X$ and $Y$, it is possible to obtain individual one-dimensional pdfs $f_X(x)$ and $F_Y(y)$, also known as marginal pdfs of $X$ and $Y$, respectively. The marginal pdfs $f_X(x)$ and $f_Y(y)$ are obtained as

$$f_X(x) = \int_{-\infty}^{+\infty} f_{X,Y}(x,y) \, dy \text{ and } f_Y(y) = \int_{-\infty}^{+\infty} f_{X,Y}(x,y) \, dx. \tag{1.33}$$

### 1.3.4 CONDITIONAL PROBABILITY DENSITY FUNCTION

We can obtain marginal probability functions using the joint density function. Sometimes it is necessary to compute the pdf of random variable $X$ given a specific value $y$ of random variable $Y$. This is called conditional pdf of $X$ given $Y$, and is denoted as $f_{X|Y}(x|y)$. The conditional pdf is defined in terms of joint and marginal pdf as

$$f_{X|Y}(x|y) = \frac{f_{X,Y}(x,y)}{f_Y(y)} \text{ and } f_{Y|X}(y|x) = \frac{f_{X,Y}(x,y)}{f_X(x)} \text{ or} \tag{1.34}$$

$$f_{X,Y}(x,y) = f_{X|Y}(x|y) f_Y(y) = f_{Y|X}(y|x) f_X(x). \tag{1.35}$$

The function $f_{X|Y}(x|y)$ is a function of variable $x$ with fixed $y$. The conditional probability density function satisfies all the properties of an ordinary pdf, i.e.,

$$f_{X|Y}(x|y) \ge 0 \text{ and } \int_{-\infty}^{+\infty} f_{X|Y}(x|y) \, dx = 1. \tag{1.36}$$

# Fundamentals of Probability Theory

## 1.3.5 STATISTICAL INDEPENDENCE

The set of $n$ random variables $X_1, X_2, \ldots, X_n$ are statistically independent if and only if (iff), the n-dimensional joint pdf can be factored into the product as

$$f_{X_1, X_2, \ldots, X_n}(x_1, x_2, \ldots, x_n) = \prod_{i=1}^{n} f_{X_i}(x_i). \tag{1.37}$$

Hence, we can say that two random variables are statistically independent if

$$f_{X,Y}(x,y) = f_X(x) f_Y(y). \tag{1.38}$$

Similarly, three random variables $X$, $Y$, and $Z$ are independent if

$$f_{X,Y,Z}(x,y,z) = f_X(x) f_Y(y) f_Z(z). \tag{1.39}$$

We can also express the statistical independence in terms of conditional pdf. Using Eq. (1.35) and Eq. (1.38), we can write

$$f_{X|Y}(x|y) f_Y(y) = f_X(x) f_Y(y) \text{ or} \tag{1.40}$$

$$f_{X|Y}(x|y) = f_X(x). \tag{1.41}$$

Similarly, we have

$$f_{Y|X}(y|x) = f_Y(y). \tag{1.42}$$

The Eq. (1.41) and Eq. (1.42) represent alternate expressions for independence of two random variables $X$ and $Y$.

## 1.3.6 MOMENTS OF A RANDOM VARIABLE

The **moments** of a random variable or its distribution are the expected values of powers or related functions of a random variable. The $n^{\text{th}}$ **moment** of a random variable $X$ is defined as

$$m_n = E\{X^n\} = \int_{-\infty}^{+\infty} x^n f_X(x) \, dx. \tag{1.43}$$

The $n^{\text{th}}$ **central moment** of a random variable is defined as

$$\mu_n = E\{(X-\mu)^n\} = \int_{-\infty}^{+\infty} (X-\mu)^n f_X(x) \, dx. \tag{1.44}$$

In particular, the first moment is called the **mean** or **expected value** of a random variable and is defined as

$$E\{X\} = \int_{-\infty}^{+\infty} x f_X(x) \, dx. \tag{1.45}$$

The mean value is denoted by $\mu$ or $\mu_x$ in this book. Note that, Eq. (1.45) is obtained by substituting $n = 1$ in Eq. (1.43).

For **discrete type** random variable, the integral in Eq. (1.45) can be replaced by a sum. Suppose the random variable $X$ takes a value $x_i$ with probability $p_i$, then the pdf can be written as

$$f_X(x) = \sum_i p_i \delta(x - x_i). \tag{1.46}$$

The expected value in this case is given by

$$E\{X\} = \sum_i p_i \cdot x_i \quad \text{where} \quad p_i = P\{X = x_i\}. \tag{1.47}$$

The conditional mean of the random variable $X$ given an event $A$ is given by Eq. (1.45) if we replace $f_X(x)$ by the conditional pdf $f_{X|A}(x|A)$:

$$E\{X|A\} = \int_{-\infty}^{+\infty} x f_{X|A}(x|A) dx. \tag{1.48}$$

For **discrete type** random variable the conditional mean is given by

$$E\{X|A\} = \sum_i x_i \cdot P\{X = x_i | A\}. \tag{1.49}$$

Let us form a random variable $Y$ using a random variable $X$ and a function $g(x)$ as $Y = g(X)$. Then, the **mean of function** of a random variable is given as

$$E\{Y\} = E\{g(X)\} = \int_{-\infty}^{+\infty} g(x) f_X(x) dx. \tag{1.50}$$

Note that, Eq. (1.50) also justify the expressions given in Eq. (1.43) and Eq. (1.44) for moments and central moments. Authors are encouraged to go through standard probability theory books for proof of some of the expressions given in this section.

From Eq. (1.50), it follows that

$$E\{a_1 g_1(X) + \cdots + a_n g_n(X)\} = a_1 E\{g_1(X)\} + \cdots + a_n E\{g_n(X)\}. \tag{1.51}$$

Eq. (1.51) represents that the expectation operation is **linear**. For **complex random variable** $Z = X + jY$, the expected value is defined as

$$E\{Z\} = E\{X\} + j\{Y\}. \tag{1.52}$$

The knowledge of mean alone is not sufficient to represent a pdf of a random variable. We need at least an additional parameter to measure the spread of the pdf around the mean. For the random variable $X$ having mean $\mu$, the term $X - \mu$ represents the deviation of the random variable from its mean value. Since the deviation can take positive as well as negative values, consider the term $(X - \mu)^2$. Its average

# Fundamentals of Probability Theory

The Binomial random variable $Y$ with parameters $n$ and $p$ takes the values $0, 1, 2, \ldots, n$ with

$$P\{Y = k\} = \binom{n}{k} p^k q^{n-k}, \quad p + q = 1 \quad k = 0, 1, 2, \cdots, n. \tag{1.61}$$

The probability mass function in this case is shown in Fig. 1.4a and the corresponding cdf is a staircase function, which is shown in Fig. 1.4b.

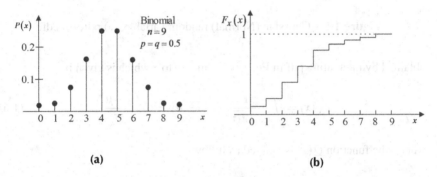

**Figure 1.4** Binomial random variable, (a) probability mass function, (b) cumulative distribution function.

### 1.3.7.2 Continuous Random Variables

The continuous random variable takes all the values in an interval which are characterized by pdf and cdf. In this section, some important continuous random variables that are used in our subsequent chapters are discussed along with their pdf and cdf.

**Gaussian (Normal) Distribution:**

Gaussian distribution is one of the most commonly used distributions. It is completely characterized by mean and variance only. The probability density function of Gaussian random variable $X$ with mean $\mu$ and variance $\sigma^2$ is given by

$$f_X(x) = \frac{1}{\sqrt{2\pi\sigma^2}} e^{-\frac{(x-\mu)^2}{2\sigma^2}}. \tag{1.62}$$

The plot for pdf is shown in Fig. 1.5a which is a bell-shaped curve symmetric around $\mu$. The Fig. 1.5a is shown for mean 0 and variance 1, i.e., $\mu = 0$ and $\sigma^2 = 1$. The constant term $\frac{1}{\sqrt{2\pi\sigma^2}}$ in Eq. (1.62) is the normalization term that maintains the area under the pdf curve to be equal to one. The cumulative distribution function can be

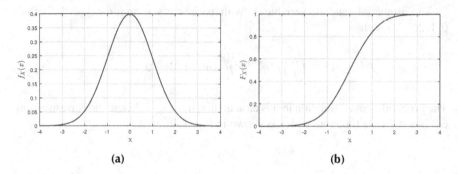

**Figure 1.5** Gaussian (Normal) random variable, (a) pdf, (b) cdf.

obtained by integrating pdf in Eq. 1.62 from $-\infty$ to $x$ which is given by

$$F_X(x) = \int_{-\infty}^{x} \frac{1}{\sqrt{2\pi\sigma^2}} e^{-\frac{(y-\mu)^2}{2\sigma^2}} dy = Q\left(\frac{x-\mu}{\sigma}\right), \quad (1.63)$$

where, the function $Q(x)$ is defined as follows

$$Q(x) = \int_{-\infty}^{x} \frac{1}{\sqrt{2\pi}} e^{-\frac{y^2}{2}} dy. \quad (1.64)$$

The Q-function is defined in detail with certain key properties in APPENDIX E. The $Q(\cdot)$ function is often available in tabulated form and it is also available as an inbuilt function in softwares like MATLAB and MATHEMATICA. Since $f_X(x)$ depends on two parameters $\mu$ and $\sigma^2$, the notation $X \sim N(\mu, \sigma^2)$ is often used to represent the pdf in Eq. (1.62). Note that, in Fig. 1.5, for the purpose of demonstration the x-axis range is chosen from $-4$ to $4$ but in theory this range will extend from $-\infty$ to $+\infty$.

**Exponential Distribution:**

The probability density function of exponentially distributed random variable with parameter $\lambda$ is given by

$$f_X(x) = \begin{cases} \lambda e^{-\lambda x}, & x \geq 0 \\ 0, & otherwise \end{cases} \quad (1.65)$$

where, $\lambda > 0$ is a parameter called rate parameter of the distribution.

The cdf of exponential distribution is given as

$$F_X(x) = P\{X \leq x\} = 1 - e^{-\lambda x}. \quad (1.66)$$

The plot for pdf and cdf for exponential random variable is shown in Fig. 1.6a and 1.6b, respectively, for different values of $\lambda$.

# Fundamentals of Probability Theory

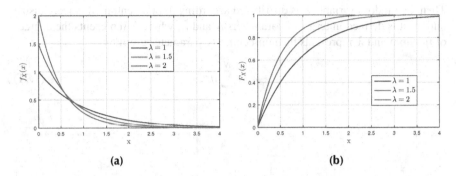

**Figure 1.6** Exponential random variable, (a) pdf, (b) cdf.

**Chi-square Distribution:**

The chi-square $(\chi_N^2)$ represents the pdf of the sum of squares of $N$ independent standard normal random variables. The probability density function of a chi-square random variable $X$ with $N$ degrees of freedom is given as

$$f_X(x) = \begin{cases} \frac{1}{2^{N/2}\Gamma(N/2)} x^{N/2-1} e^{-x/2}, & x \geq 0 \\ 0, & \text{otherwise} \end{cases} \quad (1.67)$$

where, $\Gamma(\cdot)$ represents the complete Gamma function. The respective cdf of $X$ is given by

$$F_X(x) = \frac{1}{\Gamma(N/2)} \gamma\left(\frac{N}{2}, \frac{x}{2}\right), \quad (1.68)$$

where, $\gamma(a,x)$ represents the lower incomplete Gamma function. This distribution is sometimes called the central chi-square distribution. Note that if we substitute $N = 2$ in Eq. (1.67), we obtain exponential distribution. The chi-square distribution possesses a property that the sum of independent chi-square random variables is also chi-square. Suppose, $X_i, i = 1, 2, \ldots, n$ are independent chi-square random variables each with $N_i$ degrees of freedom, respectively. Let us define a new random variable as

$$Y = X_1 + X_2 + \cdots + X_n, \quad (1.69)$$

then, the random variable $Y$ is also chi-square distributed with $N_1 + N_2 + \cdots + N_n$ degrees of freedom. For example, if $N_1 = N_2 = \cdots = N_n = N$, then $Y$ will be chi-square distributed with $nN$ degrees of freedom. The pdf and cdf of chi-square random variables are shown in Figure 1.7 for different values of $N$.

**Non-central Chi-square Distribution:**

Suppose $X_1, X_2, \ldots, X_N$ be $N$ independent, Gaussian distributed random variables with mean $\mu_i$ and unit variances. Let us form a new random variable as

$$Y = \sum_{i=1}^{N} X_i^2, \quad (1.70)$$

Then the random variable $Y$ is distributed according to non-central chi-square distribution. It is defined in terms of parameters $N$ and $\lambda$, where $N$ represents the degrees of freedom and $\lambda$ represents the non-centrality parameter given by

$$\lambda = \sum_{i=1}^{N} \mu_i^2. \tag{1.71}$$

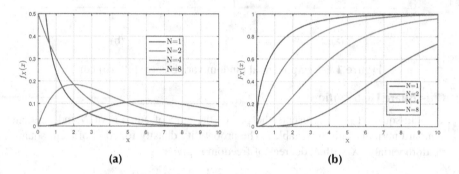

**Figure 1.7** Chi-square random variable, (a) pdf, (b) cdf.

The pdf of non-central chi-square distribution is given as

$$f_X(x) = \frac{1}{2} e^{(x+\lambda)/2} \left(\frac{x}{\lambda}\right)^{N/4-1/2} I_{N/2-1}\left(\sqrt{\lambda x}\right), \tag{1.72}$$

where $I_N(x)$ represents the modified Bessel function of first kind. The respective cdf of the non-central chi-square distribution is given as

$$F_X(x) = 1 - Q_{N/2}\left(\sqrt{\lambda}, \sqrt{x}\right), \tag{1.73}$$

where, $Q_M(a,b)$ represents the Marcum Q-function. The pdf and cdf of non-central chi-square distribution is shown in Fig. 1.8 for different values of $N$ and $\lambda$.

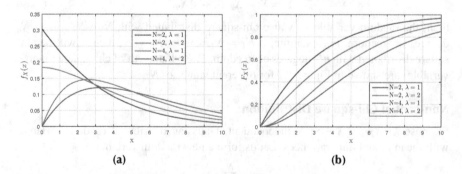

**Figure 1.8** Non-central chi-square random variable, (a) pdf, (b) cdf.

# Fundamentals of Probability Theory

**Rayleigh Distribution:**

The probability density function of Rayleigh random variable $X$ (often used in modeling the fading phenomenon occurring in wireless communication) with parameter $\sigma^2$ is given by

$$f_X(x) = \begin{cases} \frac{x}{\sigma^2} e^{-x^2/2\sigma^2}, & x \geq 0 \\ 0, & \text{otherwise} \end{cases} \quad (1.74)$$

The corresponding cumulative distribution function is given by

$$F_X(x) = 1 - e^{-x^2/2\sigma^2}. \quad (1.75)$$

The Rayleigh distribution is used to model the randomly received signal amplitude values in communication systems. The plots for pdf and cdf of Rayleigh distribution are shown in Fig. 1.9.

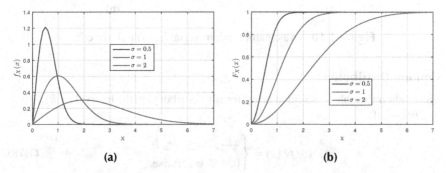

**Figure 1.9** Rayleigh random variable, (a) pdf, (b) cdf.

**Nakagami Distribution:**

Nakagami distribution is a generalization of Rayleigh distribution. This distribution is frequently used to model the fading environment in wireless communication. The pdf of this distribution is given by the following formula

$$f_X(x) = \begin{cases} \frac{2}{\Gamma(m)} \left(\frac{m}{\Omega}\right)^m x^{2m-1} e^{-mx^2/\Omega}, & x \geq 0 \\ 0, & \text{otherwise} \end{cases} \quad (1.76)$$

where, $m$ is the shape parameter and $\Omega > 0$ is used to control the spread. The corresponding cdf is given by

$$F_X(x) = \frac{\gamma\left(m, \frac{m}{\Omega} x^2\right)}{\Gamma(m)}, \quad (1.77)$$

where, $\Gamma(a,x)$ represents the lower incomplete Gamma function.

The parameter *m* controls the shape of the pdf of Nakagami distribution. Nakagami distribution provides greater flexibility compared to Rayleigh distribution in modeling randomly fluctuating fading channels in wireless communication. Note that substituting $m = 1$ results in Rayleigh distribution, and the parameter *m* controls the tail distribution. The pdf and cdf of this distribution for different parameter settings are shown in Fig. 1.10.

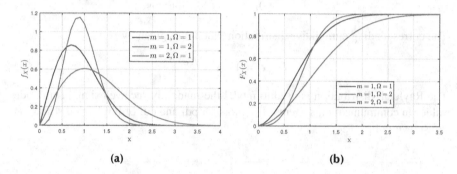

**Figure 1.10** Nakagami random variable, (a) pdf, (b) cdf.

**Uniform Distribution:**

The random variable $X$ is said to be uniformly distributed in the interval $(a, b)$, $-\infty < a < b < \infty$, if

$$f_X(x) = \begin{cases} \frac{1}{b-a}, & a \leq x \leq b \\ 0, & \text{otherwise} \end{cases} \quad (1.78)$$

The notation $X \sim U(a, b)$ is used to denote uniform distributed random variable in the interval $(a, b)$. The cdf of the uniformly distributed random variable $X$ is given by

$$F_X(x) = \begin{cases} 1, & x \geq b \\ \frac{x-a}{b-a}, & a \leq x < b \\ 0, & x < a \end{cases} \quad (1.79)$$

The pdf and cdf of the uniform random variable $X$ for specific values of $a = -2$ and $b = 2$ are shown in Fig. 1.11. The mean (expected value) and the variance of a uniformly distributed random variable $X$ are given by

$$E(X) = \frac{a+b}{2}, \quad (1.80)$$

$$\text{Var}(X) = \frac{1}{12}(b-a)^2. \quad (1.81)$$

# Fundamentals of Probability Theory

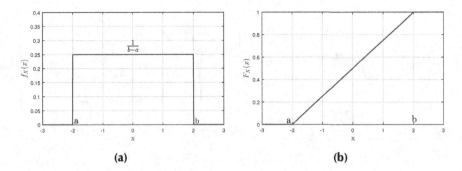

**Figure 1.11** Uniform random variable, (a) pdf, (b) cdf.

**Gamma Distribution:**

The probability density function of Gamma distributed random variable $X$ with parameters $\alpha$ and $\beta$, $\alpha, \beta > 0$, is given as

$$f_X(x) = \begin{cases} \frac{x^{\alpha-1}}{\Gamma(\alpha)\beta^\alpha} e^{-x/\beta}, & x \geq 0 \\ 0, & \text{otherwise} \end{cases} \tag{1.82}$$

where, $\Gamma(x)$ represents the Gamma function. The Gamma pdf is denoted by the notation $G(\alpha, \beta)$.

The shape of the Gamma density function depends the values of $\alpha$ and $\beta$. It takes a wide variety of shapes depending on the values of $\alpha$ and $\beta$. The pdf $f_X(x)$ is strictly decreasing for $\alpha < 1$ and $f_X(x) \to \infty$ as $x \to 0$ and $f_X(x) \to 0$ as $x \to \infty$.

The Gamma distribution is a generalized distribution that has few other distributions as its special cases. The exponential random variable is a special case of Gamma distribution with $\alpha = 1$. With $\alpha = n/2$ and $\beta = 2$, we get chi-square ($\chi^2$) distribution with $n$ degrees of freedom.

The cumulative distribution of a Gamma random variable is obtained by integrating Eq. (1.82) from 0 to $x$.

$$F_X(x) = \int_0^x \frac{y^{\alpha-1}}{\Gamma(\alpha)\beta^\alpha} e^{-y/\beta} dy. \tag{1.83}$$

Using the change of variable as $y/\beta = z$ and carrying out the integration, the cdf is obtained as

$$F_X(x) = \frac{1}{\Gamma(\alpha)} \gamma(\alpha, \beta x). \tag{1.84}$$

The plots for pdf and cdf of Gamma random variable are shown in Fig. 1.12 for different values of $\alpha$ and $\beta$.

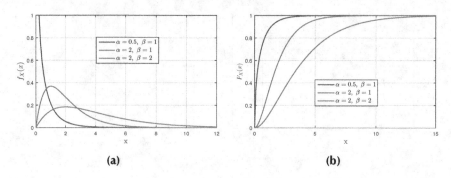

**Figure 1.12** Gamma random variable, (a) pdf, (b) cdf.

### 1.3.8 THE MARKOV AND CHEBYSCHEV INEQUALITIES

In general, the knowledge of mean and variance is insufficient to provide enough information to determine the distribution of a random variable. However, for the random variable $X$, we can obtain certain bounds for probabilities of the form $P\{|X| \geq a\}$. For nonnegative random variable $X$ with mean $E[X]$, the **Markov inequality** states that

$$P\{X \geq a\} \leq \frac{E\{X\}}{a}. \tag{1.85}$$

The proof for Eq. (1.85) inequality is given in APPENDIX A.1.

Let the mean of a random variable $X$ be $E\{X\} = \mu_X$ and the variance be $Var\{X\} = \sigma_X^2$, and we are interested in bounding the probability $P\{|X - \mu_X| \geq a\}$. The Chebyshev inequality states that

$$P\{|X - \mu_X| \geq a\} \leq \frac{\sigma_X^2}{a^2}. \tag{1.86}$$

Chebyshev inequality can be obtained from Markov inequality. Let $C^2 = (X - \mu_X)^2$, then the Markov inequality applied to $C^2$ gives

$$P\{C^2 \geq a^2\} \leq \frac{E\{(X - \mu_X)^2\}}{a^2} = \frac{\sigma_X^2}{a^2}. \tag{1.87}$$

Note that, the terms $C^2 \geq a^2$ and $\{|X - \mu_X| \geq a\}$ represent equivalent events. Using this in Eq. (1.87) gives Chebyshev inequality given in Eq. (1.86).

### 1.3.9 THE SAMPLE MEAN AND THE LAWS OF LARGE NUMBERS

Let $X$ be a random variable with true mean as $\mu_X$, which is unknown. Suppose $X_1, X_2, \cdots, X_n$ denote $n$ independent, repeated measurements of $X$. We note that the

# Fundamentals of Probability Theory

$X_i's$ are independent and identically distributed (iid) random variables having the same pdf as $X$. We use **sample mean** to estimate the expected value of $E\{X\}$ as

$$M_n = \frac{1}{n}\sum_{i=1}^{n} X_i. \tag{1.88}$$

Note that $M_n$ itself is a random variable and we will compute the mean value and variance of $M_n$ to check the effectiveness Eq. (1.88) to estimate the expected value of $X$, that is, $E\{X\}$. We also investigate the behavior of $M_n$ as $n$ becomes large.

A good estimator should give the correct value of the parameter being estimated, that is, $E\{M_n\} = \mu$, and it should not vary too much about the correct value of this parameter, that is, $E\left\{(M_n - \mu)^2\right\}$ should be small.

We can compute the expected value of $M_n$ as

$$E\{M_n\} = E\left\{\frac{1}{n}\sum_{i=1}^{n} X_i\right\} = \frac{1}{n}\sum_{i=1}^{n} E\{X_i\} = \mu. \tag{1.89}$$

Since $X_i's$ are iid having mean $\mu$. Hence, on an average, the sample mean is equal to $E\{X\} = \mu$, and we say that the sample mean represents unbiased estimator of $\mu$.

Now, the variance of $M_n$ can be computed as

$$E\left\{(M_n - \mu)^2\right\} = E\left\{(M_n - E\{M_n\})^2\right\}. \tag{1.90}$$

With $S_n = X_1 + X_2 + \cdots + X_n$, we can write $M_n = S_n/n$. Since $X_i's$ are iid, the variance of $S_n$ is given by $Var\{S_n\} = n \times Var\{X_i\} = n \times \sigma_X^2$. Using this we have

$$Var\{M_n\} = \frac{1}{n^2} Var\{S_n\} = \frac{n\sigma_X^2}{n^2} = \frac{\sigma_X^2}{n}. \tag{1.91}$$

The Eq. (1.91) states that the variance of the sample mean tends to zero as the number of samples is increased. This implies that the probability that the sample mean is close to the true mean approaches one as $n$ becomes very large. The Chebyshev inequality can be utilized to formalize this statement as

$$P\{|M_n - E\{M_n\}| \geq \varepsilon\} \leq \frac{Var\{M_n\}}{\varepsilon^2}, \tag{1.92}$$

where, $\varepsilon > 0$. Using Eq. (1.89) and Eq. (1.91) in Eq. (1.92), we get

$$P\{|M_n - \mu| \geq \varepsilon\} \leq \frac{\sigma_X^2}{n\varepsilon^2}. \tag{1.93}$$

If we consider the compliment of the event in Eq. (1.93), we obtain

$$P\{|M_n - \mu| < \varepsilon\} \geq 1 - \frac{\sigma_X^2}{n\varepsilon^2}. \tag{1.94}$$

Hence, we can choose any error $\varepsilon$ and probability $1-\delta$ and select the number of samples $n$ so that $M_n$ is within $\varepsilon$ distance of the mean $\mu$ with probability $1-\delta$ or greater.

If we let $n$ approach infinity in Eq. (1.94), we get

$$\lim_{n\to\infty} P\{|M_n - \mu| < \varepsilon\} = 1. \tag{1.95}$$

#### 1.3.9.1 Weak Law of Large Numbers

Let $X_1, X_2, \ldots$ be a sequence of independent and identically distributed random variables with finite mean $E\{X\} = \mu$ and finite variance, then for $\varepsilon > 0$,

$$\lim_{n\to\infty} P\{|M_n - \mu| < \varepsilon\} = 1. \tag{1.96}$$

It states that for a large enough fixed value of $n$, the sample mean using $n$ samples will be close to the true mean with high probability.

#### 1.3.9.2 Strong Law of Large Numbers

Let $X_1, X_2, \ldots$ be a sequence of independent and identically distributed random variables with finite mean $E\{X\} = \mu$ and finite variance, then

$$P\left\{\lim_{n\to\infty} M_n = \mu\right\} = 1. \tag{1.97}$$

The law of large number states that with probability 1, every sequence of the sample mean calculations will eventually approach and stay close to true mean.

### 1.3.10 CENTRAL LIMIT THEOREM (CLT)

Suppose we have a sequence of independent and identically distributed (iid) random variables $X_1, X_2, \ldots$ having mean $\mu$ and variance $\sigma^2$. Let us define the sum of first $n$ random variables in the sequence as $S_n$,

$$S_n = X_1 + X_2 + \cdots + X_n. \tag{1.98}$$

The central limit theorem states that, for large value of $n$, the cdf of properly normalized $S_n$ approaches that of a Gaussian random variable. Because of this, we can approximate cdf of $S_n$ with that of a Gaussian random variable. The central limit theorem is the reason why the Gaussian random variable appears in so many diverse applications.

We know since $X_1, X_2, \ldots$ are iid, the mean and the variance of $S_n$ is given by $n\mu$ and $n\sigma^2$, respectively. Let $Z_n$ be the zero mean unit variance random variable obtained from $S_n$ as

$$Z_n = \frac{S_n - n\mu}{\sigma\sqrt{n}}, \tag{1.99}$$

# Fundamentals of Probability Theory

then according to CLT,

$$\lim_{n\to\infty} Z_n = \frac{1}{\sqrt{2\pi}} \int_{-\infty}^{z} e^{-x^2/2} dx. \tag{1.100}$$

The special property of the central limit theorem is that the $X_i's$ can take any distribution having finite mean and variance. The proof for central limit theorem is discussed in APPENDIX A.2.

## 1.4 STOCHASTIC PROCESS

In the previous section, we discussed about random variable which deals with the random nature of the physical quantity and enables us to understand the probabilistic description of the numerical value of a random quantity. In reality, for certain random experiments, the outcomes are a function of time or space as an additional parameter. For example, speech signals are functions of time, and image data are functions of space. Hence, to understand such signals probabilistically, we have to develop a mathematical tool for the characterization of random signals. The mathematical model of probability theory that incorporates an additional parameter to characterize such a random phenomenon is called the theory of random or stochastic process. In this section, we will develop theory of stochastic process by considering that the random signals (random phenomenon) are functions of time.

### 1.4.1 DEFINITION OF STOCHASTIC PROCESS

Consider the sample space $S$ with sample points $\zeta_1, \zeta_2, \ldots, \zeta_n, \ldots$. To every $\zeta_i \in S$, let us assign a real valued function of time, $X(\zeta_i, t)$, which is also denoted as $X_i(t)$. This is demonstrated using Fig. 1.13 which shows a sample space $S$ with six sample points and corresponding six waveforms, labeled as $X_i(t), i = 1, 2, \ldots, 6$, with probabilities as

$$P\{\zeta_i\} = p_i, \quad i = 1, 2, \ldots, 6 \text{ where } 0 \leq p \leq 1 \text{ and} \tag{1.101}$$

$$\sum_{i=1}^{6} p_i = 1. \tag{1.102}$$

This indicates that the $i$th waveform has probability of $p_i$. Let us now consider some time instant say $t = t_1$ as shown in Fig. 1.13. Each outcome $\zeta_i$ of sample space $S$ has associated with it a number $X_i(t_1)$ and a probability $p_i$. Hence, the collection of numbers $X_i(t_1)$, $i = 1, 2, \ldots, 6$ forms a random variable. If the waveforms are observed at some different time instant say $t_2$, yields a different set of numbers and hence a different random variable. We can observe that at every time instant, we get a different set of six numbers leading to different random variables. This situation can be extended to the case of an infinite number of sample points in the sample space $S$. In that case, the number of waveforms associated with the outcomes is correspondingly large. This collection of time functions is called **ensemble**.

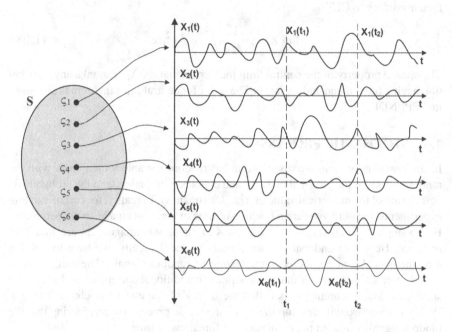

**Figure 1.13** A simple illustration of random process.

**The formal definition of random process:** A probability system composed of a sample space, an ensemble of time functions, and a probability measure is called a random process and is denoted by $X(t)$. A random process is a function of two variables $\zeta_i \in S$ and a time $t \in (-\infty, \infty)$. Hence, the random process should be denoted by $X(\zeta, t)$, but for simplicity the notation $X(t)$ is used. The function $X(\zeta, t)$ for fixed $\zeta$ is called a **realization** or a **sample path** or a **sample function**. For the given random process $X(t)$, we can make the following interpretations:

- $X(t)$ is an ensemble of functions $X(\zeta, t)$ where both $\zeta$ and $t$ are variables.
- If we fix $\zeta$ and let $t$ be a variable, then we get a single time function or a sample of a given process.
- If we fix $t$ and let $\zeta$ be a variable, then we get a random variable obtained by observing the process at a fixed time $t$.
- If we fix both $\zeta$ and $t$, then we get a real number.

### 1.4.2 STATISTICS OF STOCHASTIC PROCESS

In the stochastic process, for each value of time $t$, we have a random variable, and hence there are a noncountable infinite number of random variables. Then at time $t$,

# Fundamentals of Probability Theory

$X(t)$ is a random variable for specific value of $t$ having cdf

$$F_{X(t)}(x,t) = P\{X(t) \leq x\}. \tag{1.103}$$

The function given in Eq. (1.103) depends on the time instant $t$, and is equals to the probability of the event $X(t) \leq x$ which consists of all the outcomes $\zeta$ such that the samples of $X(t)$ at fixed time $t$ do not exceed the number $x$. Although, we have earlier used $F_X(x)$ to represent the cdf, for simplicity, we avoid subscript $X(t)$ to represent the cdf and pdf in the subsequent sections. The function $F(x,t)$ is called the first-order distribution of the random process $X(t)$. If we differentiate Eq. (1.103) with respect to $x$ we get the first order cdf of $X(t)$ as

$$f(x,t) = \frac{\partial F(x,t)}{\partial x}. \tag{1.104}$$

The joint distribution of random variable $X(t_1)$ and $X(t_2)$ obtained by sampling the process $X(t)$ at time instants $t_1$ and $t_2$ is termed as the second-order distribution of $X(t)$

$$F(x_1,x_2;t_1,t_2) = P\{X(t_1) \leq x_1, X(t_2) \leq x_2\}. \tag{1.105}$$

The corresponding density function is obtained as

$$f(x_1,x_2;t_1,t_2) = \frac{\partial^2 F(x_1,x_2;t_1,t_2)}{\partial x_1 \partial x_2}. \tag{1.106}$$

The $n^{th}$-order cdf of $X(t)$ obtained by considering $n$ time instants is given as the joint cdf of random variables $X(t_1), X(t_2), \ldots, X(t_n)$ which can be denoted as $F(x_1,x_2,\ldots,x_n;t_1,t_2,\ldots,t_n)$.

**Second-Order Properties:**

The knowledge of the function $F(x_1,x_2,\ldots,x_n;t_1,t_2,\ldots,t_n)$ is needed for every $t_i$ and $n$ in order to determine the statistical properties of the random process $X(t)$. However, in many applications, only certain averages like expected value of $X(t)$ and $X^2(t)$ are used. These quantities are defined as follows:

The **mean** at time denoted as $t$ $\eta(t)$ of $X(t)$ is the expected value of the random variable $X(t)$ given by

$$\eta(t) = E\{X(t)\} = \int_{-\infty}^{\infty} xf(x,t)\,dx. \tag{1.107}$$

Sometimes the **autocorrelation** function denoted by $R(t_1,t_2)$ of the random process $X(t)$ is obtained as the expected value of the product $X(t_1)X(t_2)$

$$R(t_1,t_2) = X(t_1)X(t_2) = \int_{-\infty}^{\infty}\int_{-\infty}^{\infty} x_1x_2 f(x_1,x_2;t_1,t_2)\,dx_1 dx_2. \tag{1.108}$$

The autocorrelation function is also denoted by $R_{xx}(t_1, t_2)$ or $R_x(t_1, t_2)$. The value of autocorrelation for $t_1 = t_2 = t$ represents the average power of the random process $X(t)$ at time $t$ and this is a function of time unless we make some assumption about the type of random process.

$$R(t,t) = E\{X^2(t)\}. \tag{1.109}$$

Similar to autocorrelation, the **autocovariance** of the process represents the covariance of the random variables at $t_1$ and $t_2$,

$$C(t_1, t_2) = E\{[X(t_1) - \eta(t_1)) X(t_2 - \eta(t_2))]\} \tag{1.110}$$
$$= R(t_1, t_2) - \eta(t_1)\eta(t_2). \tag{1.111}$$

The value of autocovariance for $t_1 = t_2 = t$, i.e., $C(t,t)$, represents the variance of random process $X(t)$ at time $t$.

### 1.4.3 STATIONARITY

A random process $X(t)$ is called **strict-sense stationary (SSS)** if its statistical properties are invariant to a shift of the origin. Let the random process $X(t)$ be observed at time instants $t_1, t_2, \ldots, t_n$ and $\bar{X}(t)$ be the corresponding random vector. Then, the random process $X(t)$ is said to be strict sense stationary if the joint pdf $f_{\bar{X}(t)}(\bar{x})$ is invariant to the translation of the time origin, that is,

$$f_{\bar{X}(t+T)}(\bar{x}) = f_{\bar{X}(t)}(\bar{x}), \tag{1.112}$$

where, $(t+T) = (t_1 + T, t_2 + T, \ldots, t_n + T)$. For $X(t)$ to be stationary, Eq. (1.112) should be valid for any finite set of time instants $t_i$, $i = 1, 2, \ldots, n$ and every time shift $T$ and a dummy vector $\bar{x}$. The random process $X(t)$ is called nonstationary process if it is not stationary.

A random process $X(t)$ is called **wide sense stationary** or stationary in wide sense if it satisfies the following properties

- The mean of random process $X(t)$ is constant, i.e., independent of time,

$$E\{X(t)\} = \eta_X(t) = \eta_X. \tag{1.113}$$

- The autocorrelation of a random process $X(t)$ depends only on the time different, i.e., $\tau = t_k - t_j$

$$R_X(t_k, t_j) = R_X(t_k - t_j). \tag{1.114}$$

The wide sense stationary represents the weak kind of stationarity. All the processes that are stationary in the strict sense are also wide sense stationary, but the reverse is not true.

### 1.4.3.1 Properties of Autocorrelation Function

The autocorrelation of the wide sense stationary process satisfies certain properties. The autocorrelation function of the wide sense stationary process is defined as

$$R_X(\tau) = E\{X(t+\tau)X(t)\}, \tag{1.115}$$

where, $\tau$ represents the time difference between the two time instants $t_k$ and $t_j$ at which the random process $X(t)$ is observed.

- The mean square value of random process $X(t)$ is $R_X(0) = R_X(\tau)|_{\tau=0}$. The proof of this follows from Eq. (1.115) as

$$R_X(0) = E\{X^2(t)\}. \tag{1.116}$$

Note that $R_X(0)$ is constant and hence for WSS process, mean and mean square values are independent of time.

- The autocorrelation function is an even function of $\tau$; that is

$$R_X(\tau) = R_X(-\tau). \tag{1.117}$$

The proof for this property follows from the fact that $E\{X(t+\tau)X(t)\} = E\{X(t)X(t+\tau)\}$. Since the process is wide sense stationary, we have

$$R_X[(t+\tau)-t] = R_X[t-(t+\tau)].$$
$$R_X(\tau) = R_X(-\tau). \tag{1.118}$$

- The autocorrelation function is maximum at origin,

$$R_X(0) \geq R_X(\tau). \tag{1.119}$$

To prove this property consider the quantity $E\{[X(t) \pm X(t+\tau)]^2\}$. This represents an expectation of squared quantity and hence is nonnegative. Therefore expanding $E\{[X(t) \pm X(t+\tau)]^2\}$, we get

$$E\{X^2(t) \pm 2X(t)X(t+\tau) + X^2(t+\tau)\} \geq 0. \tag{1.120}$$

Since expectation is a linear operator, we can take expectation inside the bracket which gives us

$$E\{X^2(t)\} \pm 2E\{X(t)X(t+\tau)\} + E\{X^2(t+\tau)\} \geq 0,$$
$$2R_X(0) \pm 2R_X(\tau) \geq 0,$$
$$R_X(\tau) \pm R_X(\tau) \geq 0,$$
$$R_X(0) \geq |R_X(\tau)|. \tag{1.121}$$

Note that, $E\{X^2(t+\tau)\} = E\{X(t+\tau)X(t+\tau)\} = R_X(t+\tau-t-\tau) = R_X(0)$.

- If the sample functions of the random process $X(t)$ are periodic with period $T_0$, then the autocorrelation function is also periodic with the same period.
  In this case, if we consider the autocorrelation function $R_X(\tau) = E\{X(t+\tau)X(t)\}$ for $\tau \geq T_0$, the sample function repeats with period $T_0$, so does the product $X(t+\tau)X(t)$ and the expectation of this product which is the autocorrelation also repeats with the same period.

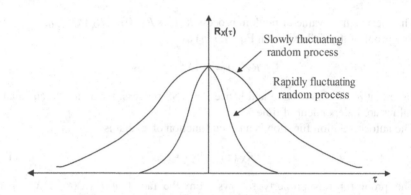

**Figure 1.14** Autocorrelation function of slowly and rapidly fluctuating random process.

The autocorrelation function $R(\tau)$ provides a means to determine the dependence of two random variables obtained by observing the random process at two time instants separated by the distance $\tau$. For a zero mean random process, $R(\tau)/R(0)$ is the correlation coefficient of random variables obtained by observing the random process at time instants $\tau$ seconds apart. As shown in Fig. 1.14, the value of $R_X(\tau)$ decreases rapidly from maximum value $R_X(0)$ for rapidly fluctuating random process and its value decays slow for slowly fluctuating random process.

### 1.4.4 RANDOM PROCESSES THROUGH LINEAR SYSTEM

We know the relationship between input and output of linear time invariant (LTI) system when the input to the system is deterministic. In that case, the output of the system is obtained by convolving the input signal with the impulse response of the system, i.e., the impulse response completely characterizes an LTI system. To deal with real life environment, the response of deterministic systems when excited by random signals needs to be studied. In that regard, we will develop a relationship between input and output when an LTI system is excited by random signals.

Consider the scenario shown in Fig. 1.15 where the random process $X(t)$ is the input to an LTI system having impulse response $h(t)$ (known and deterministic) to produce the output random process $Y(t)$.

# Fundamentals of Probability Theory

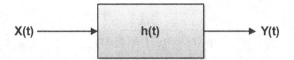

**Figure 1.15** Transmission of random process through linear time invariant system.

We will try to characterize $Y(t)$ in terms of $X(t)$ and $h(t)$. Specifically, we evaluate the expected value of random process $\eta_Y(t)$ and the autocorrelation function $R_Y(t_1, t_2)$ when $X(t)$ is wide sense stationary.

Let $x_i(t)$ be $i^{th}$ sample function of random process $X(t)$ is applied as an input to an LTI system producing the corresponding output as $y_i(t)$. Then $y_i(t)$ represents the sample function of output process $Y(t)$. Since the system is LTI, we can write $y_i(t)$ as the convolution of $x_i(t)$ and $h(t)$, that is,

$$y_i(t) = \int_{-\infty}^{\infty} h(\tau) x_i(t-\tau) d\tau. \quad (1.122)$$

Since the relationship given in Eq. (1.122) is true for every realization of $X(t)$, we can write

$$Y(t) = \int_{-\infty}^{\infty} h(\tau) X(t-\tau) d\tau. \quad (1.123)$$

Hence, when the random process $X(t)$ is applied, we get another random process $Y(t)$ at the output. Let us compute the mean of the output process $Y(t)$, i.e.,

$$\eta_Y(t) = E\{Y(t)\} = E\left\{\int_{-\infty}^{\infty} h(\tau) X(t-\tau) d\tau\right\}. \quad (1.124)$$

We can interchange the order of expectation and the integration provided that the expected value of $X(t)$ is finite for every value of $t$, and the system is stable. After interchanging expectation and integration, we get

$$\eta_Y(t) = \int_{-\infty}^{\infty} h(\tau) E\{X(t-\tau)\} d\tau. \quad (1.125)$$

Since we have assumed that $h(\tau)$ as deterministic, expectation operation is applied on random process only. If we assume that $X(t)$ is wide sense stationary, then $E\{X(t-\tau)\}$ is a constant, since we know that for wide sense stationary process

$E\{X(t)\} = \eta_X$. Using this in Eq. (1.125), we have

$$\eta_Y(t) = \eta_X \int_{-\infty}^{+\infty} h(\tau) d\tau,$$
$$= \eta_X H(0), \qquad (1.126)$$

where, $H(0) = H(f)|_{f=0}$ and $H(f)$ is the transfer function of the given LTI system $h(t)$. The transfer function is obtained by taking the Fourier transform of the impulse response. Note that the expected value of the output process $\eta_Y(t)$ is constant.

Let us now compute the autocorrelation function of the output process $Y(t)$ at two time instants $t_1$ and $t_2$,

$$R_Y(t_1, t_2) = E\{Y(t_1) Y(t_2)\},$$
$$= E\left\{ \int_{-\infty}^{\infty} h(\tau_1) X(t_1 - \tau_1) d\tau_1 \int_{-\infty}^{\infty} h(\tau_2) X(t_2 - \tau_2) d\tau_2 \right\}. \qquad (1.127)$$

Interchanging the order of integration and expectation we have

$$R_Y(t_1, t_2) = \int_{-\infty}^{\infty} \int_{-\infty}^{\infty} h(\tau_1) h(\tau_2) E\{X(t_1 - \tau_1) X(t_2 - \tau_2)\} d\tau_1 d\tau_2,$$
$$= \int_{-\infty}^{\infty} \int_{-\infty}^{\infty} h(\tau_1) h(\tau_2) R_X(t_1 - t_2 - \tau_1 + \tau_2) d\tau_1 d\tau_2,$$
$$= \int_{-\infty}^{\infty} \int_{-\infty}^{\infty} h(\tau_1) h(\tau_2) R_X(\tau - \tau_1 + \tau_2) d\tau_1 d\tau_2, \qquad (1.128)$$

where $\tau = t_1 - t_2$. We can see that the autocorrelation depends only on $\tau$. Since the expected value $E\{Y(t)\}$ of output process $Y$ is constant and the autocorrelation depends only on the time difference $\tau$, we conclude that the process $Y(t)$ is wide sense stationary. This is one of the key results which states that when an LTI system is excited by a wide sense stationary process, then the output of the system is also wide sense stationary.

### 1.4.5 POWER SPECTRAL DENSITY (PSD)

Fourier transform can be defined for a deterministic signal. However, the same cannot be used on random signals. If one has to do it then the Fourier transform has to be applied on every sample function that complicates the frequency domain representation for random signals. Instead one can use the concept of power spectral density when we are dealing with the wide sense stationary random process.

# Fundamentals of Probability Theory

The power spectral density of wide sense stationary process $X(t)$ is defined as the Fourier transform of its autocorrelation function $R_X(\tau)$,

$$S_X(f) = \int_{-\infty}^{\infty} R_X(\tau) e^{-j2\pi f \tau} d\tau. \tag{1.129}$$

The autocorrelation function $R_X(\tau)$ can be obtained by taking inverse Fourier transform of $S_X(f)$ as

$$R_X(\tau) = \int_{-\infty}^{\infty} S_X(f) e^{j2\pi f \tau} df. \tag{1.130}$$

As evident from Eq. (1.129) and Eq. (1.130), power spectral density $S_X(f)$ and the autocorrelation function $R_X(\tau)$ form a Fourier transform pair. Eq. (1.129) and Eq. (1.130) are popularly known as the Weiner-Khinchine relations.

### 1.4.5.1 Properties of Power Spectral Density

Following are some of the key properties of power spectral density.

1. The zero frequency value of the power spectral density for a WSS process represents the total area under the autocorrelation function curve, that is,

$$S_X(0) = \int_{-\infty}^{\infty} R_X(\tau) d\tau. \tag{1.131}$$

Eq. (1.131) follows directly from Eq. (1.129) by substituting $f = 0$.

2. The total area under the curve of power spectral density of a WSS process equals the mean square value of process,

$$E\{X^2(t)\} = \int_{-\infty}^{\infty} S_X(f) df. \tag{1.132}$$

This property follows from Eq. (1.130) and noting that

$$R_X(0) = E\{X^2(t)\} = \int_{-\infty}^{\infty} S_X(f) df. \tag{1.133}$$

3. The power spectral density is an even function of frequency, that is, $S_X(f)$ is real and $S_X(f) = S_X(-f)$. This property is due the property of autocorrelation function, that is, $R_X(\tau)$ is real and even function of $\tau$ and the property of Fourier transform that says the Fourier transform of real and even function is also real and even.

4. The power spectral density of wide sense stationary process is always non-negative.

$$S_X(f) \geq 0 \text{ for all } f. \tag{1.134}$$

### 1.4.5.2 Output Spectral Density of an LTI System

The power spectral density of output of an LTI system for WSS input is obtained by taking Fourier transform of $R_Y(\tau)$ we have derived in Eq. (1.128),

$$S_Y(f) = \int_{-\infty}^{\infty} R_Y(\tau) d\tau,$$

$$= \int_{-\infty}^{\infty}\int_{-\infty}^{\infty}\int_{-\infty}^{\infty} h(\tau_1) h(\tau_2) R_X(\tau + \tau_2 - \tau_1) e^{-j\pi f \tau} d\tau_1 d\tau_2 d\tau. \quad (1.135)$$

Using the change of variable as $u = \tau + \tau_2 - \tau_1$ we have

$$S_Y(f) = \int_{-\infty}^{\infty}\int_{-\infty}^{\infty}\int_{-\infty}^{\infty} h(\tau_1) h(\tau_2) R_X(u) e^{-j\pi f(u-\tau_2+\tau_1)} d\tau_1 d\tau_2 du,$$

$$= \int_{-\infty}^{\infty} h(\tau_2) e^{j2\pi f \tau_2} d\tau_2 \int_{-\infty}^{\infty} h(\tau_1) e^{-j2\pi f \tau_1} d\tau_2 \int_{-\infty}^{\infty} R_X(u) e^{-j2\pi f u} du,$$

$$= H^*(f) H(f) S_X(f),$$

$$= |H(f)|^2 S_X(f). \quad (1.136)$$

Eq. (1.136) relates the input and output power spectral densities to the transfer function of LTI system. Note that $R_Y(\tau)$ can also be found by computing first Eq. (1.136) to obtain $S_Y(f)$ and then taking inverse Fourier transform.

### 1.4.6 GAUSSIAN RANDOM PROCESS

There are certain random processes such as the Gaussian process, Poisson process, Markov process, etc, which are important to deal with real life phenomenon. In this book, only the Gaussian process is discussed because of its great practical and mathematical significance in communication theory. Gaussian random process is significant in practice because often the noise process encountered in communication can be modeled as Gaussian processes. In addition to that, it has a neat mathematical structure, which makes the mathematical analysis quite feasible.

Let the random variables of the process $X(t)$ be observed at time instants $t_1, t_2, \ldots, t_n$. Let us denote the random variables obtained at these time instants as $X(t_1) = X_1, X(t_2) = X_2, \ldots, X(t_n) = X_n$ and corresponding random vector as $\bar{X} = (X_1, X_2, \ldots, X_n)$. The process $X(t)$ is called Gaussian random process, if $f_{\bar{X}}(\bar{x})$ is an n-dimensional joint Gaussian pdf for every $n \geq 1$ and $(t_1, t_2, \ldots, t_n) \in (-\infty, \infty)$. The n-dimensional Gaussian pdf is given as

$$f_{\bar{X}}(\bar{x}) = \frac{1}{(2\pi)^{n/2} |C_{\bar{X}}|^{1/2}} exp\left[-\frac{1}{2}(\bar{x} - \eta_{\bar{x}})^T C_{\bar{X}}^{-1}(\bar{x} - \eta_{\bar{x}})\right], \quad (1.137)$$

where, $C_{\bar{X}}$ is the covariance matrix, $|\cdot|$ denotes the determinant, $\eta_{\bar{X}}$ is the mean vector and superscript $T$ represents the transpose of a matrix. The covariance matrix $C_{\bar{X}}$ is given by

$$C_{\bar{X}} = \begin{bmatrix} COV(X_1,X_1) & COV(X_1,X_2) & \cdots & COV(X_1,X_n) \\ COV(X_2,X_1) & COV(X_2,X_2) & \cdots & COV(X_2,X_n) \\ \vdots & \vdots & & \vdots \\ COV(X_n,X_1) & COV(X_n,X_2) & \cdots & COV(X_n,X_n) \end{bmatrix}, \quad (1.138)$$

where, $COV(X_i,X_j) = E\{(X_i - \eta_{X_i})(X_j - \eta_{X_j})\}$.

Note that the knowledge of mean and covariance matrix are sufficient to represent the n-dimensional Gaussian pdf. If the random process is WSS, that is, the mean value of the process is constant and covariance $COV[X(t_i)X(t_j)]$ depends only on $t_i - t_j$, then the pdf is independent of time origin. In other words, the wide sense stationary Gaussian process is also a strict sense stationary.

## 1.4.7 WHITE NOISE

Any noise process with a flat power spectral density, i.e., having power of all the frequency components in equal proportion for $-\infty < f < \infty$ is called white noise. The word "white" is used in analogy with white light, which contains all colors in equal proportion. The power spectral density of a white noise process $W(t)$ is denoted as,

$$S_W(f) = \frac{N_0}{2}, \text{ Watts/Hz}, \quad (1.139)$$

where, the fraction $1/2$ is introduced to indicate that the half power is associated with the positive frequencies and another half with the negative frequencies. Here, $N_0$ is a constant.

If the noise process is Gaussian and has flat power spectral density, we call it **White Gaussian Noise (WGN)**. Note that white and Gaussian are two different attributes of a random process. It is not necessary for white noise to be Gaussian or the Gaussian noise to be always white. Only when "whiteness" together with "Gaussianity" exists, the process is quantified as a white Gaussian noise process. Also, note that the white process by definition is a zero mean process.

As we have PSD of white noise as $S_W(f) = N_0/2$, taking the inverse Fourier transform, the autocorrelation function of white noise process is given by

$$R_W(\tau) = \frac{N_0}{2}\delta(\tau). \quad (1.140)$$

The power spectral density and the autocorrelation function for white noise are pictorially represented in Fig. 1.16. Fig. 1.16b and Eq. (1.140) implies that any two samples of white noise, no matter how closely they are observed, are uncorrelated. That is,

$$R_W(\tau) = 0, \text{ for } \tau \neq 0. \quad (1.141)$$

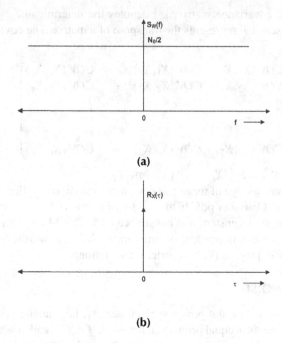

**Figure 1.16** White noise, (a) power spectral density, (b) autocorrelation function.

The Gaussian random variable has the property that if two Gaussian random variables are uncorrelated, they are also independent. Hence, if the noise process happens to be Gaussian, then any two white Gaussian noise samples are statistically independent.

# 2 Introduction

Wireless communication systems utilize the radio frequency (RF) spectrum as their propagation environment. Radio frequency spectrum is the range of electromagnetic frequencies from approximately 30 kHz to 300 GHz that can be used for radio communications. Radio frequency spectrum is an essential component for wireless communication infrastructure. The need for RF spectrum is increasing due to the rapid growth in users, applications and bandwidth requirements of modern wireless communication systems. Traditionally, the national regulatory body or the government allocates spectrum bands exclusively to the wireless applications by providing license to use those frequency bands. Most of the available RF spectrum has already been allocated to the existing wireless systems. This causes only a small part of it can be licensed to new wireless applications. A study by the Spectrum Policy Task Force (SPTF) of the Federal Communications Commission (FCC) has showed that the RF spectrum is significantly underutilized due to the static spectrum allocation and dedicated access through licensing resulting in the wastage of the resources [64, 65]. Spectrum holes or spectrum opportunities arise due to the current static spectrum licensing scheme. Spectrum holes are defined as frequency bands which are allocated to the licensed users but are not utilized in some locations at some times and therefore can be used by the unlicensed users or secondary users (SUs) [88].

The International Telecommunication Union (ITU), European Union telecom regulatory bodies, FCC and OFCOM (UK Office of Communications) have recognized that the optimal use of radio spectrum is dependent on flexible spectrum management policies and the multi-time sharing of this precious resource. Looking into spectrum scarcity and underutilized existing licensed spectrum, many countries have allocated some spectrum for unlicensed use. Few of the prominent unlicensed frequency bands in US are tabulated in TABLE 2.1. Recently, the FCC proposed making up to 1200 MHz of spectrum available for use by unlicensed devices in the 6 GHz band, or five times the current spectrum available. FCC is also considering to make spectrum above 95 GHz available for unlicensed use. It is being proposed to add a new experimental license type that would permit experimental use on any frequency from 95 GHz to 3 THz with no limits on geography or technology. The current status of unlicensed spectrum bands in USA can be accessed from official website of FCC. In UK, many spectrum bands are allowed for unlicensed use which serves a variety of applications. Some of the prominent unlicensed bands in UK correspond to 2.4 GHz, 5.1 GHz, 5.5 GHz and 60 GHz. The detailed lists of current license exempt bands in UK with their applications are listed in TABLE 2.2. This information is obtained from official website of the UK office of communication commonly known as OFCOM [1]. It is the UK government-approved regulatory and competition authority for the broadcasting, telecommunications and postal industries of the United Kingdom. In the TABLES 2.1 and 2.2, only major frequency bands are listed. Other license bands exempted are not tabulated here.

**Table 2.1**
**List of Few Unlicensed Frequency Bands in US [2]**

| Band | Frequencies (MHz) |
|---|---|
| ISM/Spread Spectrum | $902 - 928, 2400 - 2483.5, 5725 - 5850$ |
| Unlicensed Personal Communication Services (PCS) | $1910 - 1930, 2390 - 2400$ |
| Millimeter Wave | $59,000 - 64,000$ |
| Unlicensed National Information Infrastructure (U-NII) | $5150 - 5350, 5725 - 5825$ |
| Millimeter Wave (Expansion) | $57,000 - 59,000$ |

Presently in many developing countries like India, a large part of the RF spectrum is controlled by the government, with only a minimal amount of frequencies being allocated for unlicensed use. However policy makers are beginning to recognize the importance of allocating more unlicensed spectrum. The recent trend is towards allocating more and more band for unlicensed access in order to improve the spectrum utilization and at the same time allowing emerging wireless technologies to evolve. The unlicensed devices need to dynamically access these unlicensed bands for their use. These unlicensed devices are commonly referred to as cognitive radio (CR) in literature.

## 2.1 COGNITIVE RADIO

Cognitive Radio is a new paradigm of designing wireless communications systems which aims to maximize the utilization of the underutilized RF spectrum. The term cognitive radio was first introduced by Joseph Mitola III in his papers in 1999 [135, 136, 137]. Simon Haykin in [88] defines the cognitive radio, built on the software defined radio, as an intelligent wireless communication system that is aware of its environment and uses the methodology of understanding-by-building to learn from the environment and adapt to statistical variations in the input stimuli, with two primary objectives in mind: (1) highly reliable communication whenever and wherever needed and (2) efficient utilization of radio spectrum. CR users, i.e., secondary users (SUs), are allowed to access the licensed spectrum bands of primary users (PUs), i.e., licensed users, as long as they do not cause unacceptable interference to the PUs. CR is a promising solution to the continuously increasing RF spectrum demands and the spectrum scarcity caused by the fixed frequency allocations [64, 65]. This resulted in CR technology gaining increased attention and has been highlighted by both standards and regulatory bodies [52, 66, 153]. The pictorial representation of group of secondary CR network coexisting with a primary network is shown in Fig 2.1.

## Table 2.2
## Unlicensed Frequency Bands in UK [1]

| Application | Frequencies |
|---|---|
| Active Medical Implants, Inductive Applications, Non-Specific Short Range Devices, Citizens Band Radio, Railway Applications | $9KHz - 37.5MHz$ |
| Telemetry, short range devices and model control, alarms, cordless telephone | $40MHz$ |
| Non-specific short range devices | $49MHz$ |
| Assistive listening devices, non-specific short range devices, alarms, railway applications | $173MHz$ |
| Active medical implants | $401 - 406MHz$ |
| Alarms, telemetry, telecommand and medical and biological applications | $458MHz$ |
| White space devices | $470 - 790MHz$ |
| Non-specific Short Range Devices, Radio Microphones, Wireless audio/multimedia, Cordless Telephone, RFID, Transport and Traffic Telematics | $863 - 876MHz$ |
| Non-Specific Short Range Devices, Assistive Listening Devices, RFID | $915 - 921MHz$ |
| Short Range Devices, Radio Determination, Wireless audio/multimedia, Industrial/Commercial Telemetry and Tele-command, Medical implants | $2.4 - 2.5GHz$ |
| Radio Determination, Radar, Wireless Access Systems | $4.5 - 5.725GHz$ |
| Radio Determination, Radar, Non-Specific Short Range Devices, Short Range Indoor Data Links, Wireless Video Cameras | $5.725 - 7.1GHz$ |
| Radar, Radio Determination, Transport and Traffic Telematics | $21.65 - 27.7GHz$ |
| Non-Specific Short Range Devices, Radar, Wideband Data Transmission Devices, Transport and Traffic Telematics | $57 - 71GHz$ |
| Radar, Transport and Traffic Telematics | $75 - 85GHz$ |
| Non-Specific Short Range Devices | $122 - 123GHz, 244 - 246GHz$ |

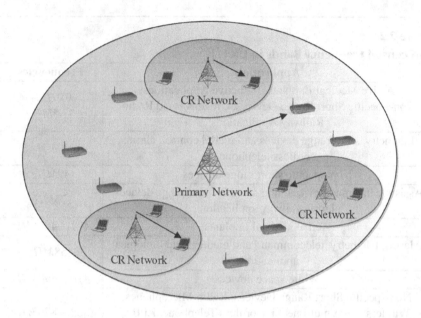

**Figure 2.1** Cognitive radio networks existing within a primary network.

There are three possible models for Cognitive Radio implementations: the interweave, the underlay, and the overlay models.

- **Interweave:** In this model, the secondary users or unlicensed users are forbidden to use the band licensed to the primary or licensed users. The secondary users are allowed to opportunistically access the licensed spectrum only when the primary users are absent. To do that, the secondary users have to dynamically identify the unused licensed spectrum, also known as spectrum holes, at a particular time and geographical location. The process of identifying the unused radio spectrum is termed as spectrum sensing, which is one of the primary enabling technology for the interweave model. Once the primary user starts transmitting the licensed band again, the secondary users need to quickly vacate the licensed band to reduce the interference to the primary user. In summary, this model works on the interference free basis and utilizes the spectrum holes.
- **Underlay:** In this model, the primary and secondary users can coexist with each other, and hence the network is termed as a spectrum sharing network [23, 77, 174]. In this model, both the primary user and the secondary user can transmit over the same radio spectrum [78, 175, 203]. Hence, there is no need for spectrum sensing to detect spectrum holes. The primary users always have higher priority over the use of spectrum than the secondary users. Besides, the spectrum sharing must be maintained in such a way that the interference experienced by the primary users is below a predefined threshold. This predefined interference threshold is a

term in literature as an interference temperature. The unlicensed secondary users may have to limit the transmit power, cancel interference, and implement guard regions around the primary receivers to reduce the interference [106, 107]. Due to this interference constraint, the underlying technology finds its application mainly in short-range communications.
- **Overlay:** In overlay model, concurrent transmission from the primary user and secondary users is allowed. It is assumed that the primary data sequence is known in advance to the secondary transmitter to negate the effect of interference [174]. Also, this knowledge can be utilized by secondary user nodes to cooperate with the primary network by relaying PU message.

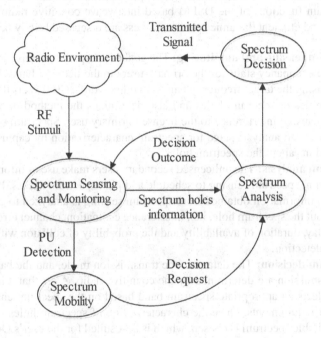

**Figure 2.2** Dynamic Spectrum Management Framework for interweave network model [5, 131].

The spectrum measurements by FCC show that the occupancy below 3 GHz is between 15 – 85 % [64, 65]. The interweave cognitive radio model can provide sufficient reliability and reasonable guaranteed quality of service (QoS), and because of that, industrial standardization bodies have preferred the interweave cognitive radio model over the other two. As a result, standards such as IEEE 802.11 and IEEE 082.11af have been proposed that make use of the interweave model. Because of the popularity of the interweave model and the standardization efforts by the standardization bodies, this book is primarily focused on the enabling technologies for the interweave cognitive radio model. Many essential components, such as spectrum sensing, spectrum analysis, spectrum decision, and spectrum mobility, are needed to realize this model. However, our primary focus in this book is limited to spectrum sensing, which is the most important feature of the interweave model. The other aspects of cognitive radio are not discussed in detail here.

The concept of interweave cognitive radio model aims at improving the spectrum utilization by reusing the under-utilized radio resources in an opportunistic manner [5, 88, 108]. A complete architecture for the model with the detailed functionality is provided by the dynamic spectrum management framework (DSMF) proposed in [131]. The DSFM is required for interweave cognitive radio technology to realize efficient spectrum utilization. Fig. 2.2 shows four main blocks of DSMF: spectrum sensing and monitoring, spectrum analysis, spectrum decision and spectrum mobility. The main functions of the DSFM based interweave cognitive radio to support intelligent and efficient dynamic spectrum access are discussed briefly below:

- **Spectrum sensing and monitoring:** The goal of the spectrum sensing is to determine the occupancy status of the primary users in the licensed bands by periodically sensing the target frequency bands. In other words, CR detects the spectrum opportunities or spectrum holes and also determines the method of accessing it without causing interference to the licensed primary user. Spectrum sensing also helps spectrum analysis stage for spectrum characterization by capturing proper observations about the spectrum holes.
- **Spectrum analysis:** The unlicensed secondary users make use of information obtained from spectrum sensing to schedule and plan spectrum access. In spectrum analysis, information obtained from spectrum sensing is analyzed to gain knowledge about the spectrum hole, i.e., interference estimation, channel error rate, link layer delay, duration of availability and the probability of collision with a PU due to miss detection.
- **Spectrum decision:** The data rate, the transmission mode, and the bandwidth of the transmission are determined by the cognitive radio. After that, the cognitive radio selects an appropriate spectrum band based on the spectrum characteristic and user requirements. Once the characterization of spectrum holes is done, the best available spectrum is chosen, which is best suited for the user's QoS requirements. The secondary users access the spectrum holes once a decision is made on spectrum access based on spectrum analysis. Spectrum access is performed based on a cognitive media access control (MAC) protocol, which aims to avoid interference with the primary users and other secondary users. The CR transmitter is also required to perform negotiation with the CR receiver to synchronize the transmission so that effective communication can be achieved successfully.
- **Spectrum mobility:** It is the function related to the change of frequency band of CR users. When a PU starts transmitting in the licensed band currently utilized by the secondary user, the secondary users have to change to an idle spectrum band. The task of spectrum mobility is to suspend the secondary transmission, vacate the channel, and resume the ongoing communication using another unoccupied primary channel. This change in the operating frequency band is

referred to as spectrum handoff, which is a key element of spectrum mobility. There are two main approaches for handoff: reactive [204, 205, 207] and proactive [172, 207, 214, 236]. The reactive approach assumes that the secondary user finds the target backup channel using reactive spectrum sensing. Whereas, in the proactive approach, it is assumed that the secondary users have sufficient knowledge of the primary user activity so that it can predict the arrival of the PU and then evacuates the channel before the PU arrival.

One of the main functionalities of CR is the spectrum sensing [31, 87]. In this book, our main focus is on spectrum sensing. The book covers the study of existing narrowband spectrum sensing techniques considering practical scenario and also presents novel detection algorithms for wideband spectrum sensing to improve the detection performance. One of our approaches also uses cooperative spectrum sensing for wideband spectrum sensing. In what fallows, we provide quick introduction to the spectrum sensing where we discuss narrowband, wideband and cooperative spectrum sensing.

## 2.2 SPECTRUM SENSING

One of the crucial requirements of SU is to monitor the usage activity in the licensed spectrum to exploit underutilized spectrum (referred to as the spectrum opportunity or spectrum holes) without causing harmful interference to the PUs. Furthermore, PUs do not have any obligation to share and change their operating parameters for sharing spectrum with SUs. Hence, SU should be able to independently detect spectrum holes without any help from PUs; this ability is called spectrum sensing, which is considered as one of the critical components in cognitive radio networks [182]. Before utilizing the licensed spectrum, SUs need to identify whether the band is occupied by any PU. During the use of a particular licensed band, SUs need to continuously monitor whether any PU has become active in that band and if so, they need to vacate that band. To achieve this, we need efficient spectrum sensing techniques that minimize interference to the PU and at the same time maximize the spectrum utilization. There are mainly two types of spectrum sensing, namely, narrowband and wideband sensing. Excellent survey on spectrum sensing can be found in [11]. In the following sections, we give brief introduction to both narrowband and wideband sensing.

### 2.2.1 NARROWBAND SPECTRUM SENSING

Narrowband spectrum sensing is used for finding the occupancy status of a single PU licensed band. The term narrowband implies that the bandwidth of the signal is sufficiently small such that the channel frequency response can be considered flat. In other words, the bandwidth of our interest is smaller than the coherence bandwidth that represents the maximum bandwidth over which the channel response is flat. The task of narrowband spectrum sensing is to decide whether the narrow slice of PU spectrum is available for secondary usage or not. Basically, the task of narrowband spectrum sensing is to decide between two hypotheses $H_0$ and $H_1$ defined as

$$\begin{cases} H_0 : y(n) = w(n), \\ H_1 : y((n)) = x(n) + w(n). \end{cases} \quad (2.1)$$

Here, $y(n)$ represents the received signal at the secondary user, $x(n)$ is the PU signal observed at the SU, $w(n)$ represents the circularly symmetric complex Additive white Gaussian noise (AWGN) with mean zero and variance $\sigma_w^2$ such that $w(n) \sim \mathscr{CN}(0, \sigma_w^2)$, and $n$ represents the time index.

The performance of the narrowband spectrum sensing algorithms are generally measured in terms of two probabilities: the probability of detection ($P_D$) and the probability of false alarm ($P_F$) [20, 59, 60, 199]. $P_D$ represents the probability of correctly detecting the primary user signal when it is actually present. On the other hand, $P_F$ is the probability that the detector wrongly detects the primary user as present when it is actually not. Ideally, we want $P_D$ as 1 and the $P_F$ as 0, which is not possible under the low signal to noise ratio (SNR) regime. An attempt is made to minimize $P_F$ so that spectrum utilization can be improved and $P_D$ is maximized to reduce interference to the PU. The general block diagram of the narrowband spectrum sensing architecture is shown in Fig. 2.3 where $F_s$ represents the sampling frequency used in ADC. One of the fundamental elements of RF chain in narrowband spectrum sensing is the bandpass filter, which filters out the narrowband of interest for detection.

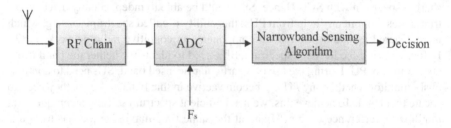

**Figure 2.3** Basic block diagram of narrowband sensing architecture at the secondary user.

We now give a brief explanation on some of the most popular narrowband spectrum sensing techniques. The detailed survey on narrowband spectrum sensing can be found in [124, 224].

### 2.2.1.1 Matched Filter Detection

The matched filter (MF) [31, 99, 101, 239] based spectrum sensing represents an optimal detection scheme that maximizes the output SNR in the presence of additive white Gaussian noise. It is also known as an optimal coherent detector [99]. In

MF detection, the received signal is correlated with the template for detecting the presence of the known signal in the received signal. Whenever the correlation peak appears, the detector assumes that the PU signal is present; otherwise, the channel is assumed to be free [28]. Here, the CR requires prior knowledge about PUs signal. It is also required that CRs be equipped with carrier synchronization and timing devices, leading to increased implementation complexity. Besides, it is not always possible to achieve good performance with MF because of practical implementation issues. For example, with limited knowledge about the frequency offset, fading channel, and proper timing, the correlation is significantly affected, resulting in degraded performance.

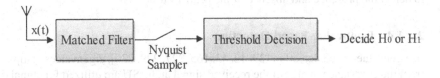

**Figure 2.4** Basic block diagram of matched-filter-based spectrum sensing.

### 2.2.1.2 Cyclostationary Detection

Generally, the human-made signals are nonstationary. Some of them are cyclostationary, i.e., their statistics exhibit periodicity, which may be caused by modulation and coding or even be intentionally produced to aid channel estimation and synchronization or intentionally induced to assist spectrum sensing [129, 186, 187]. In this detection scheme, the periodicity of the statistic such as mean and autocorrelation of the primary user signal is utilized for detection of the random signal with a particular modulation type in a background of noise and other modulated signals [6, 45, 46, 73, 103, 120, 158]. This detection scheme is capable of differentiating the primary signal from the interference and noise. An ability to differentiate between different primary systems adds to the advantages of this detector.

The detector computes cyclic autocorrelation function $R_y(\alpha, \tau)$ given as

$$R_y(\alpha, \tau) = \frac{1}{M} \sum_{n=1}^{M} E[y(n)y^*(n+\tau)] e^{-j\alpha n}, \qquad (2.2)$$

where, $y(t)$ is the received signal, $E[\cdot]$ represents the expectation operation, $y^*(n)$ represents the complex conjugate of $y(n)$, $\alpha$ is the cyclic frequency, and $\tau$ represents the time lag value. The cyclic autocorrelation $R_y(\alpha, \tau)$ computed in Eq. (2.2) acts as a decision statistic which is compared to a threshold in order to decide whether primary user is active or not. This technique requires partial prior information about the primary signal and its computational cost is relatively high. Furthermore, the sensing interval for the cyclostationary detectors is known to be high [121].

### 2.2.1.3 Covariance-Based Detection

The primary signal received at the SU is usually correlated due to the dispersive channel, use of multiple antennas, or even over sampling. Hence, the covariance of the received signal when PU is transmitting is different from when it is not transmitting. This property is used to differentiate between presence and the absence of the primary signal [228, 231]. Here, the covariance matrix of the received signal is computed which is different for signal and the noise. The structure of the covariance matrix of the noise is known at the receiver. Based on this property, the covariance matrix is converted into another matrix. The off-diagonal elements of the resultant matrix are zero when there is no transmission and few are nonzero when there is PU transmission. This property of the transformed covariance matrix is used to detect the presence and absence of the primary user.

### 2.2.1.4 Eigenvalue-Based Detection

In eigenvalue-based detection [29, 42, 43, 165, 227, 229, 230], eigenvalues computed from the covariance matrix of the received signal at the SU are utilized for signal detection. The ratio of maximum or average eigenvalue to the minimum eigenvalue is used to detect presence of the primary user signal. Detection schemes based on covariance matrix and eigenvalues represent blind detection schemes and they do not require any information about the PU signal, channel and noise variance. The eigenvalue-based detection techniques are independent of the effects of noise uncertainty and perform better when the signals to be detected are highly correlated.

### 2.2.1.5 Energy Detection

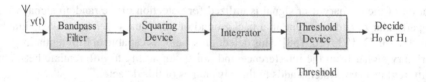

**Figure 2.5** Basic block diagram of energy-detection-based spectrum sensing.

Energy detection (ED) [59, 60, 105, 199] is a non-coherent detection method that do not require any prior knowledge about the PUs. Fig. 2.5 shows the block diagram of energy detection (ED) based method for spectrum sensing. The received signal is passed through the bandpass filter, which filters out the band of interest. This filtered signal is then passed through the squaring device, followed by an integrator. The output of the integrator gives the energy, i.e., decision statistic, which is compared with the threshold to decide the active/ inactive, i.e., presence or absence, status of the primary user. Note that Fig. 2.5 may not be the only way to represent energy detection.

# Introduction

It is also possible to sample the bandpass filtered signal and square and add the absolute values of samples to compute the decision statistic. In this case $y(t)$ is converted into its samples $y(n)$. This decision statistic is then compared with the threshold to decide the presence or the absence of the primary user signal. The computational complexity of ED-based spectrum sensing is relatively low. However, ED performs poorly under the low SNR scenario and when noise uncertainty is considered. Even then, due to its simplicity and reduced computational complexity, ED represents a very popular spectrum sensing technique although other techniques may exhibit better performance. Considering the advantages of ED, we adopt it for our study in this book and propose different algorithms based on the same.

## 2.2.2 WIDEBAND SPECTRUM SENSING

In wideband spectrum sensing, multiple bands are sensed for finding spectrum opportunities. Typical example plot of a wideband spectrum is shown in Fig. 2.6. In general terms, the wideband consists of a large number of narrow subbands. Here, we aim to sense a frequency bandwidth that exceeds the coherence bandwidth of a channel. For example, wideband spectrum sensing aims at finding spectrum opportunities over the whole ultra-high frequency (UHF) TV band (between 300 MHz to 3 GHz). To do this, one should note that narrowband sensing techniques cannot be directly used for the wideband case. This is because these techniques make a single binary decision for the whole spectrum and thus cannot identify individual spectrum holes that lie within the broad frequency spectrum. The wideband spectrum sensing can be broadly classified into two classes based on the sampling frequency used. They correspond to Nyquist wideband spectrum sensing and sub-Nyquist wideband spectrum sensing. A detailed survey on wideband spectrum sensing can be found in [182].

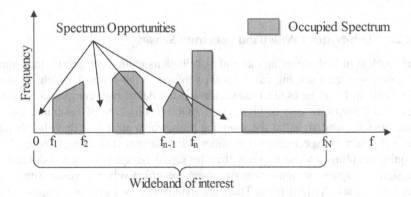

**Figure 2.6** Wideband spectrum consisting of $N$ frequency bands.

## 2.2.2.1 Nyquist Wideband Spectrum Sensing

In Nyquist wideband spectrum sensing [63, 154, 155, 185, 193], we directly acquire the wideband signal using a standard analog to digital converter (ADC) which samples the wideband signal at the Nyquist sampling rate. This approach assumes that it is feasible to sample the wide spectrum at the conventional Nyquist rate. We then use signal processing techniques to detect the spectrum opportunities. The general block diagram for possible implementation of Nyquist wideband spectrum sensing is shown in Fig. 2.7. The sampling rate required in this approach is very high and hence it is practically unaffordable. Due to high sampling rate, the computational complexity of this approach is very high. Also, the sensing time that is the time required to make the decision is comparatively high. Due to these reasons, sensing the wideband at the Nyquist rate poses a significant challenge in terms of building the necessary hardware that operates at a sufficiently high sampling rate and in designing high-speed signal processing algorithms.

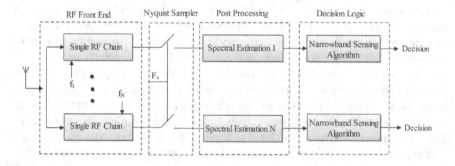

**Figure 2.7** Basic block diagram of Nyquist-sampling-based wideband spectrum sensing.

## 2.2.2.2 Sub-Nyquist Wideband Spectrum Sensing

The problem of high sampling rate and high implementation complexity in Nyquist wideband spectrum sensing can be solved by using sub-Nyquist wideband spectrum sensing. It can be used to reduce the sensing delay, high computational complexity and hardware cost of high speed ADC. In this type of sampling, we acquire the wideband signal at the sampling rates which are lower than the Nyquist rate and detect the spectrum opportunities using these partial measurements. Sub-Nyquist sampling scheme requires that the signal be sparse in some domain. If the signal is sparse in some domain, it can be effectively recovered from samples taken at sub-Nyquist rates. There are mainly two types of sub-Nyquist wideband spectrum sensing schemes and they correspond to compressive-sensing-based [81, 85, 194, 195, 197, 210] and the one that uses multichannel sub-Nyquist wideband sensing [133, 179, 200]. One of the possible compressed sensing based

# Introduction

sub-Nyquist spectrum sensing implementation is shown in Fig. 2.8. Sub-Nyquist wideband spectrum sensing is an extensively researched area and many possible implementations are available in the literature. In this book, our discussion on wideband spectrum sensing is limited to Nyquist wideband spectrum sensing only. The readers are encouraged to go through the literature in order to gain better insights into this topic.

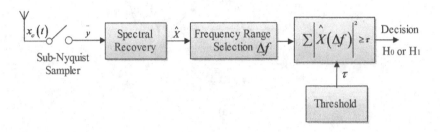

**Figure 2.8** Compressive-sensing-based sub-Nyquist wideband spectrum sensing using spectral domain energy detection approach.

## 2.2.3 COOPERATIVE SPECTRUM SENSING

Spectrum sensing is an important functionality of CR to improve the spectrum utilization and at the same time limiting the harmful interference to the licensed users. However, the detection performance in practice is greatly affected by the effect of multipath fading and shadowing. To combat the effect of these issues, cooperative spectrum sensing (CSS) has been shown to be effective method to improve the detection performance by exploiting the spatial diversity. In CSS, multiple secondary users known as cooperating secondary users (CSUs) collaborate by sharing their information in order to detect the spectrum opportunities. CSS is divided into three categories based on how the CSUs share their information in the network: centralized [74, 198, 202], distributed [115] and relay-assisted [68, 69, 237]. The schematic for these three scheme are shown in Fig. 2.9. As shown in Fig. 2.9a for the centralized CSS, the CSUs report their information using reporting channels to the fusion center (FC) which then takes the decision on the presence or the absence of the PU. In distributed CSS, the CSUs do not rely on the FC to make the decision instead they share information among themselves and collectively make the decision as shown in Fig. 2.9b. Finally, in relay assisted CSS shown in Fig. 2.9c, some of the CSUs serve as a relay to assist in forwarding sensing results from other CSUs. There are mainly two types of data sharing possible which are soft combining and hard combining. In soft combining, CSUs share actual decision statistic among themselves whereas in hard combining final decision is shared. The CSS can be used for both narrowband and wideband sensing. The detailed survey on cooperative spectrum sensing can be found in [7] where the issues such as cooperation methods, cooperative gain and the cooperation overhead are all discussed.

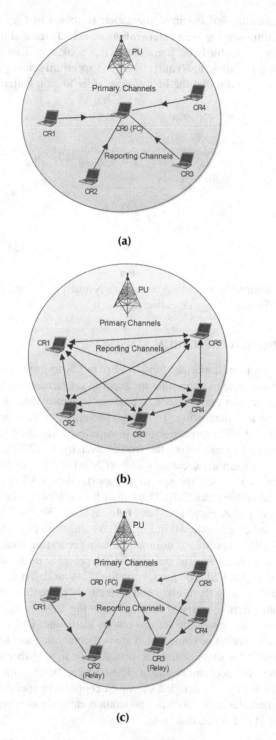

**Figure 2.9** Classification of cooperative spectrum sensing, (a) centralized, (b) distributed, and (c) relay-assisted.

## 2.2.4 MACHINE-LEARNING-BASED SPECTRUM SENSING

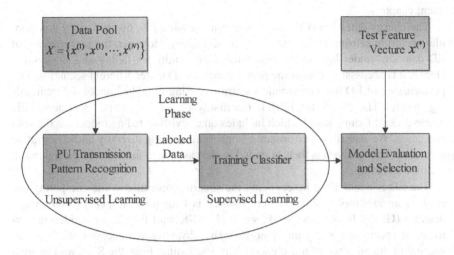

**Figure 2.10** Block diagram of machine-learning-based spectrum sensing approach.

Deep learning is a subset of machine learning that enables a computer to solve more complex problems. Deep learning techniques have shown excellent performance in areas such as computer vision, audio and speech, natural language processing, etc [21, 39, 48, 61, 94, 100, 148, 159, 162, 166, 167, 195, 201, 209, 221, 222, 223, 234]. Researchers have recently starting exploring the use of machine learning for next-generation wireless communication [80, 91, 92, 93, 147, 161, 190, 208, 218, 233]. Spectrum sensing is a binary classification task where the secondary users have to classify the primary user's presence or absence. Hence, machine learning techniques can be utilized for such a classification. Recently few works have reported the use of deep learning for spectrum sensing [54, 89, 109, 111, 118, 215]. There are mainly two types of machine learning techniques: supervised and unsupervised. Supervised learning makes use of labeled data for learning, whereas, in unsupervised learning, unlabeled data is used. The block diagram in Fig. 2.10 depicts the machine learning-based spectrum sensing approach. First, the unlabeled dataset consisting of received wireless signals is used to create a labeled dataset using unsupervised learning. This labeled data set is then used to train a classifier. The classifier model is then evaluated, and the best model is selected. The trained classifier is then used to classify the new incoming data. Note that the machine learning technique based spectrum sensing is out of the scope of this book and hence it is not discussed in full depth. Readers are encouraged to go through referred papers to get more insights into this subtopic.

## 2.3 BOOK CONTRIBUTIONS

Having provided brief introduction to spectrum sensing, we now summarize the important contributions of this book, the details of which are discussed in the subsequent chapters.

The energy detection (ED) based spectrum sensing [59, 60, 105, 199] is a popular spectrum sensing technique due to its simplicity. However, the performance of ED degrades under low SNR, noise uncertainty, multipath fading and shadowing. Hence, it is necessary to study the performance of ED under different scenarios. The performance of ED for narrowband spectrum sensing is studied under different fading channels [18, 19, 59, 60, 170]. In our first work, we give the performance of ED under general fading model which includes other existing fading models as the special cases. We also show performance improvement using diversity and cooperative detection. This analysis is then extended to the case when there exists shadowing in addition to fading.

The ED is generalized by changing the squaring operation while computing energy by an arbitrary positive power $p$ which is known as the generalized energy detector (GED). In our next work, we study SNR wall for GED considering narrowband spectrum sensing under no diversity, diversity and cooperative spectrum sensing in the presence of noise uncertainty and fading. First, the SNR wall expressions are obtained considering AWGN channel and the conclusions are drawn based on them. The effects of diversity on the SNR walls are discussed. The analysis is then extended to channel with Nakagami fading where the SNR walls are obtained numerically. The effect of fading on SNR wall is discussed. All the obtained results for SNR walls are validated using Monte Carlo simulations.

In the literature, use of diversity for performance improvement is mostly limited to narrowband spectrum sensing. To this end, we propose the use of diversity for wideband spectrum sensing in our next work. We make use of square law combining (SLC) and square law selection (SLS) diversity schemes in our proposed algorithms. The performance improvement is shown against the existing algorithm. Also in existing literature, the theoretical analysis for wideband spectrum sensing is limited to no fading. In our work, we give theoretical analysis considering Nakagami fading channel and also validate it using Monte Carlo simulations. The effects of different parameters on the performance of the proposed algorithms are also discussed.

Finally, in our last work, we propose novel detection algorithms for cooperative wideband spectrum sensing. The proposed algorithm use hard and soft combining for cooperation. We use hard combining because it incurs reduced cooperation overhead. We provide complete theoretical analysis of the proposed algorithm and validate it using Monte Carlo simulation. We show that the proposed algorithm outperforms the existing algorithm used without cooperation. Also, it performs better than our previously proposed algorithms that make use of diversity by choosing appropriate number of cooperating secondary users. A soft combining based algorithm performs even better than the hard combining based algorithm. However, the cooperation overhead is very high in soft combining based algorithm when compared to the hard combining based algorithm.

To summarize, in this book, we have addressed the problem of narrowband as well as wideband spectrum sensing. This includes the following works:

- Performance of energy detection (ED) based spectrum sensing for narrowband over $\eta - \lambda - \mu$ fading channel,
- Performance of generalized energy detector (GED) under diversity and cooperation by considering noise uncertainty and fading,
- Use of diversity for wideband spectrum sensing under fading,
- Detection algorithm for cooperative wideband spectrum sensing.

## 2.4 TOUR OF THE BOOK

The contents of this book are organized as follows. The literature review is presented in Chapter 3. As already discussed, the performance of energy detector degrades under low SNR, fading and shadowing and hence it is important to study its performance under different fading scenarios. We analyze the performance of energy-detection-based narrowband spectrum sensing for under $\eta - \lambda - \mu$ fading channel in Chapter 4. It is a general fading model which includes other fading models as the special cases. Here, the expressions for average probability of false alarm and the average probability of detection are derived. We also show the performance improvement using diversity and collaborative detection. The analysis is then extended to the case when there exists shadowing in addition to fading.

Generally, it is assumed that the true noise variance is known. In practice, what is actually known is the expected value of the noise variance. The true value of the noise variance is not fixed at particular time as well as location giving rise to what is known as noise uncertainty. The performance of generalized energy detector (GED) is greatly affected by noise uncertainty and fading which gives rise to the phenomenon known as the SNR wall. In Chapter 5, we analyze SNR wall for GED under diversity as well as cooperation when the noise uncertainty and fading are considered.

In Chapters 4 and 5, the analysis is restricted to narrowband spectrum sensing. Next, we consider the case of wideband spectrum sensing. Since the use of diversity for performance improvement in narrowband spectrum sensing is well studied in the literature, we propose the use of diversity for wideband spectrum sensing in Chapter 6. Here, two new detection algorithms are proposed and it is shown that the performance improves when compared to the no diversity case. The complete theoretical analysis for the proposed algorithms are given under Nakagami fading channel and validated using simulations.

In Chapters 7, we propose novel detection algorithm that uses cooperative spectrum sensing with hard combining for wideband sensing. Hard combining is used for data fusion since it incurs reduced cooperation overhead. Complete theoretical analysis for the proposed algorithm is given and validated using simulation. Performance improvement is shown against no cooperation. Also, by choosing appropriate number of cooperating secondary users, we show that this algorithm performs better when compared to our algorithms that use diversity. In this chapter, we also discuss the novel detection algorithm that uses both cooperative sensing and antenna diversity for performance improvement. The soft combining is used for data sharing in this algorithm. It is demonstrated that this algorithm outperforms all the algorithms proposed in previous chapters.

Finally, in the Chapter 8, we conclude the book by summarizing the main contributions and by listing out future research directions. It may be of interest to note that much of the material discussed in this book has been published in our works [32, 33, 34, 35, 36, 37, 38, 123] and interested reader may refer the same for details.

# 3 Literature Review

In this chapter, we provide a review of the literature for spectrum sensing highlighting the insights of current research status in these areas. We first review the literature on narrowband spectrum sensing in Section 3.1 followed by wideband spectrum sensing in Section 3.2 and cooperative spectrum sensing in Section 3.3.

## 3.1 NARROWBAND SPECTRUM SENSING

Various detection techniques for narrowband spectrum sensing have been investigated, namely, matched filtering based detection [31, 101], cyclostationary detection [45, 103, 120], covariance based detection [228, 231], eigenvalue based detection [227, 229, 230] and energy detection [59, 60, 105, 199]. The optimal way to detect the occupancy status of the PU signal under AWGN is the matched filter detection [31, 101], since it maximizes the received signal to noise ratio. However, a matched filter effectively requires demodulation of primary user signal. This means that cognitive radio has a priori knowledge of primary user signal at both Physical (PHY) and Medium Access Control (MAC) layers. The main advantage of matched filter is that it does not require significant amount of time to achieve high processing gain since only $O(1/SNR)$ samples are needed to meet a given probability of detection constraint [57]. However, the main drawback of a matched filter detection is that a CR would need a dedicated receiver for every PU class. Man-made signals are generally nonstationary. However, few of these signals are cyclostationary, i.e., their statistics exhibit periodicity, which may be caused due to the use of modulation and coding or even it may intentionally introduced to aid channel estimation and synchronization. Such periodicity can be utilized for detecting a random signal with a particular modulation type in a background of noise and other modulated signals. This is called cyclostationary detection which was first introduced in [72]. The cyclostationary detection is realized by analyzing the cyclic autocorrelation function (CAF) [53] of the received signal, or, equivalently, its two-dimensional spectrum correlation function (SCF) [73], since the spectrum redundancy caused by periodicity in the modulated signal results in correlation between widely separated frequency components [73, 114]. In [45], the noise rejection property of the cyclostationary spectrum is used to perform spectrum sensing at very low SNR. Here, the spectrum sensing algorithm for IEEE 802.22 WRAN is developed. It is shown using simulation that for a probability of false alarm as 0.1 at SNR= $-25$ $dB$, the miss detection probability of 0.1 is achieved using the proposed algorithm. The cycle frequency domain profile (CDP) is used for signal detection and preprocessing for signal classification in [103]. Signal features are extracted from CDP and a Hidden Markov Model (HMM) has been used for classification. It is shown that the CDP-based detector and the HMM-based classifier can classify PU signals at low SNRs. A generalized likelihood ratio test (GLRT) for detecting the presence of cyclostationarity using multiple cyclic frequencies is proposed in [120] and the performance improvement is shown using simulations in low SNR regime. The cyclostationary detection can be used to differentiate the primary signal from the interference and noise. It works even in very low SNR region and its performance is independent of the noise uncertainty. But, the limitation of the cyclostationary detection is that, similar to matched filter, it also requires prior knowledge about the PU signal. Added to this, the computational complexity is also high when compared to energy detector. Researchers have also attempted to solve the spectrum sensing problem by using the covariance of the received signal. We know that the statistical covariance of the signal and noise are different and hence can be used to differentiate the case where the primary user's signal is present from the case where there is only noise. The spectrum sensing algorithms based on this idea are proposed in [228, 231] in which sample covariance matrix is calculated using the limited number of received signal samples. Based on the eigenvalues of the sample covariance matrix, different spectrum sensing algorithms are proposed in [227, 229, 230]. Authors in [227] propose spectrum sensing method based on the maximum eigenvalue. The spectrum sensing method based on the ratio of the maximum eigenvalue to the minimum eigenvalue is proposed in [229, 230]. The detection scheme based on the ratio of the average eigenvalue to the minimum eigenvalue is also proposed by the authors in [230]. Spectrum sensing methods based on covariance matrix and its eigenvalues do not need any kind of knowledge about the signal, channel and the noise power. These methods perform better than ED when the received signal samples are correlated. Also, the computational complexities of these approaches are high. Energy Detection (ED) based spectrum sensing proposed in [59, 60, 105, 199] is the most popular technique due to its simplicity. ED is a noncoherent blind detection technique and it does not require any prior knowledge about the PU signal. It is simple to design and implement in practice. The computational complexities of ED is significantly less when compared to other techniques. Looking at the advantages of ED, we make use of ED in all our works.

In the literature, many studies have been dedicated to the analysis of ED based spectrum sensing by considering different communication scenarios. In [199], Urkowitz derive the probability of detection ($P_D$) and probability of false alarm ($P_F$) under additive white Gaussian noise channel. Kostylev in [105], revisited the problem of ED considering the fading conditions and obtained the expressions for $P_D$ and $P_F$ under Rayleigh, Rice and Nakagami fading channels. Closed form expressions for the average probability of detection ($\bar{P}_D$) for no diversity and with diversity under Rayleigh, Rician and Nakagami fading channels are derived in [60]. Authors in [90] derive the expressions for $\bar{P}_D$ for the no diversity case as well as the maximal ratio combining (MRC) diversity case considering Nakagami-m and Rician fadings by using the moment generating function (MGF) and probability density function (pdf) based methods. In [17], the detection performance of ED is investigated under very low SNR levels. Here, the closed form expressions are derived for the average probability of missed detection considering Rayleigh and Nakagami−m fading channels. The performance of ED over generalized $k-\mu$ and $k-\mu$ *extreme* fading channels have been investigated in [170], where the expressions for $\bar{P}_D$ are derived for no diversity case which are then subsequently extended to square law selection (SLS) diversity and for cooperative detection. The performance of ED over wireless channels with composite multipath fading and shadowing is studied in [19]. The performance of energy detector over $\eta - \mu$ fading channel is analyzed in [18]. The performance of ED over mixture gamma distribution is studied in [10], where the expressions for average probability of detection are derived. The analysis is then extended to square law selection and square law combining diversity schemes. Recently, in [22], the performance of ED is investigated over Nakagami−$q$/ Hoyt fading channel. Authors in [30], study the effects of RF impairments such as in-phase and quadrature-phase imbalance, low-noise amplifier nonlinearities, and phase noise on the performance of ED based spectrum sensing. In [8] the performance of ED is analyzed over different composite generalized multipath/gamma fading channels, namely, $k - \mu$/ gamma, $\eta - \mu$/gamma, and $\alpha - \mu$/gamma. To approximate SNR for these channels, mixture gamma distribution is utilized. The performance of ED is analyzed over two-wave with diffuse power (TWDP) fading channels in [41], where the authors obtain novel expressions for average probability of detection under TWDP fading channels. The expressions derived are utilized to study performance of ED under moderate, severe and extreme fading conditions. The use of diversity and cooperation to mitigate these effects are also demonstrated. Authors in [220] discuss ED based spectrum sensing over $\mathscr{F}$ composite fading channel. They obtain expressions for detection probabilities and extended the same to include collaborative spectrum sensing, square law selection diversity and noise uncertainty. The receiver operating characteristics (ROC) is analyzed for different conditions such as average SNR, noise variance uncertainty, time-bandwidth product, fading, shadowing, number of diversity branches, and number of cooperating secondary users. The area under the ROC curve (AUC) is derived for different multipath fading and shadowing conditions to evaluate the performance of ED. In [9], the performance of ED based spectrum sensing is analyzed over $\alpha - \eta - k - \mu$ fading channel. The analytical expressions are derived for effective rate (ER), the average detection probability and the average area under the receiver operating characteristic curve. Although, in the literature, ED based spectrum sensing is studied for different scenarios, the scope for analyzing the performance of ED under more general fading channel still exists. In this book, to begin with, the performance of ED is discussed under $\eta - \lambda - \mu$ fading model which is more general and also includes other fading models as its special cases.

In ED, also known as conventional energy detector (CED), if the noise variance is known exactly, it is possible to detect the PU signal even at very low SNR if the sensing time is made sufficiently large [101]. In practice, the noise variance varies with time as well as location giving rise to noise uncertainty [189]. The effect of worst case noise uncertainty on CED is discussed in [173, 189]. In [189], a phenomenon called SNR wall is studied. The effect of uniformly distributed noise uncertainty on the performance of CED is studied in [82, 232]. The authors in [232] derive the expression for SNR wall for CED assuming a uniform distribution for noise uncertainty. In [130], an asymptotic analysis of noise power estimation is performed for CED. The condition for existence of SNR wall is obtained and the effect of noise variance estimation on the performance is studied. The SNR wall for cooperative spectrum sensing assuming the same SNRs and the noise uncertainties for all the cooperating secondary users is discussed in [206, 226]. The performance of energy detector under a Log-Normal approximated noise uncertainty is studied in [96] and the closed form expressions for the performance measures are obtained. The performance of CED under different types of noise uncertainty models are studied in [212]. It is shown that proper selection of threshold is the key to improve the performance of CED under the existence of noise uncertainty. Authors in [95] study the energy-detection-based spectrum sensing of digital video broadcasting-terrestrial (DVB-T) signals. By making use of pilot periodicity of DVB-T signal, a low complexity noise power estimator is proposed which works under multipath channel and does not require time and frequency synchronization. It is shown that with the use of noise power estimator, CED outperforms the pilot correlation based detection which is often used in DVB-T spectrum sensing. The impact of unknown noise variance on the performance of CED for orthogonal frequency division multiplexing (OFDM) cognitive radio is investigated in [49].

The CED is generalized by replacing the squaring operation of received signal amplitude by an arbitrary positive power $p$, which is referred to as the generalized energy detector (GED) [97, 98] or the improved energy detector [47, 168, 169] or $p$-norm detector [24, 25] or the $L_p$−norm detector [138]. In [47, 169, 171], it is shown that performance of the CED can be improved by choosing a suitable value for $p$ that depends on $P_F$, the average SNR as well as on the sample size. In [24, 25], different approximations for $P_F$ and $P_D$ are developed by considering different fading scenarios. Also, for antenna diversity reception, new detection schemes called $p$−law combining (pLC) and $p$−law selection (pLS) are proposed. In [97, 98], the performance of GED is studied under noise uncertainty where the authors in [97] show that under worst case of noise uncertainty the SNR wall is not dependent on the value of $p$. It is also shown that under the assumption of uniform distribution of noise uncertainty, the CED represents the optimum ED. The expression for SNR wall is obtained in [98] where the noise uncertainty is once again chosen as uniformly distributed. It is also shown that the SNR wall does not depend on the value of $p$. The study of the detection performance is then extended to noise uncertainty having log normal distribution and the SNR wall for the same is calculated numerically. Authors in [97, 98] have derived the SNR wall for GED under noise uncertainty considering no diversity. The performance of GED is analytically studied under fast faded channel in the presence of noise. The noise here is modeled using McLeish distribution which is suitable for non-Gaussian and Gaussian noise channel.

## 3.2 WIDEBAND SPECTRUM SENSING

A simple approach of wideband spectrum sensing (WSS) is to acquire the wideband signal by sampling at the corresponding Nyquist rate using standard analog to digital converter (ADC) and then apply digital signal processing techniques to detect the spectrum opportunities. Number of researchers have studied WSS based on Nyquist sampling. Authors in [154, 155] have proposed an optimal multiband joint detection scheme for WSS in which a bank of multiple narrowband detectors are jointly optimized to improve the aggregate opportunistic throughput of a CR system while limiting the interference to the PU. This technique suffers from issues such as power consumption and nonfeasibility of ultra-high sampling ADCs. As the extension of this, authors in [67] propose an adaptive multiband spectrum sensing algorithm. The algorithm consists of two phases: the exploration phase, where substantial portion of the available channels are eliminated according to accumulated statistics and the detection phase, where multiple spectrum opportunities are finally identified among the remaining channels. A wavelet based spectrum sensing algorithm is proposed in [193] and [62] where the power spectral density (PSD) of the wideband spectrum is modeled as a series of consecutive frequency subbands, in which the PSD is smooth within each subband but exhibits discontinuities and irregularities on the border of two neighboring subbands. The multiband spectrum sensing for cognitive radio considering hardware limitations of secondary users is discussed in [216]. It is proposed that only a small portion of the wideband is sensed by the SUs at a particular time. Two different strategies, namely, random spectrum sensing strategy (RSSS) and adaptive spectrum sensing strategy (ASSS), are proposed to select channels to be sensed at a given time. To evaluate the performance of the spectrum sensing strategies, new performance metric called spectrum sensing capability (SSC) is defined and shown using experiments that the ASSS outperforms the RSSS strategy. The wavelet transform is then used to locate the boundaries of the subbands and the occupancy status of the subbands are decided based on the PSD levels in each subband. Farhang-Boroujeny in [63] proposes a filter bank approach for wideband spectrum sensing where a bank of prototype filters is used to process the wideband signal. Here, the baseband is directly estimated by using a prototype filter and the other bands are obtained by modulating this prototype filter. In each band, the corresponding portion of the wideband is down converted to form a baseband version of that subband on which a narrowband sensing algorithm is applied. The drawback of this approach is that, due to the parallel structure of the filter bank, the implementation of this algorithm requires a large number of radio frequency components. The extension of this work can be found in [104, 116]. In general a SU may not be interested in finding all the spectrum opportunities, instead the interest lies in finding sufficient numbers of spectrum opportunities. This can be achieved if we consider only a part of the wideband spectrum for sensing. Keeping this into consideration, authors in [185] propose partial band Nyquist sampling (PBNS) which samples part of the wideband instead of entire wideband thus reducing the sampling rate. Since PBNS uses the traditional Nyquist sampling it represents the simplest wideband spectrum sensing scheme.

Due to the drawback of high sampling rate in Nyquist WSS, sub-Nyquist sensing is drawing more and more interest in recent years. Two important types of sub-Nyquist wideband sensing correspond to compressive sensing and multichannel sub-Nyquist based approaches. Compressive-sensing-based wideband spectrum sensing was first introduced in [194] in which fewer samples are used to perform the sensing. Here, the number of samples closer to the information rate, rather than the inverse of the bandwidth are used. The wideband spectrum is reconstructed and then a wavelet based edge detection is used to detect the spectrum opportunities. To improve the robustness against noise uncertainty, a cyclic feature detection based compressive sensing algorithm is proposed in [195]. Authors in [225] proposed a distributed compressive sensing based sensing algorithm for cooperative multihop cognitive radio networks in order to reduce the acquisition cost. To realize the analogue compressive sensing, an analogue-to-information converter (AIC), is proposed in [197]. A quadrature analogue-to-information converter is introduced in [217] to rapidly sense the spectrum of interest. Comparisons of various architectures in compressive sensing are presented in [51]. In addition to this, a number of wideband spectrum sensing techniques that make use of compressive sensing can also be found in [81, 85, 210]. Mishali and Eldar proposed a multichannel sub-Nyquist sampling approach known as modulated wideband converter (MWC) [133] by modifying the AIC model. Authors in [71] propose novel wideband spectrum sensing algorithm based on a modulated wideband converter and sparse Bayesian learning (SBL). It is shown using simulations that the proposed algorithm performs better than MWC based orthogonal matching pursuit method. An alternative multichannel sub-Nyquist sampling approach is the multi-coset sampling. Multi-coset sampling is equivalent to choosing some samples from a uniform grid, which can be obtained using a sampling rate $f_s$ higher than the Nyquist rate. The uniform grid is then divided into blocks of $m$ consecutive samples and in each block $v$ ($v < m$) samples are retained. Thus, it is often implemented by using $v$ sampling channels with sampling rate of $f_s/m$, with different sampling channels having different time offsets. To obtain a unique solution for the wideband spectrum from these partial measurements, the sampling pattern should be carefully designed. In [200], some sampling patterns were proved to be valid for unique signal reconstruction. The advantage of the multi-coset approach is that the sampling rate in each channel is $m$ times lower than the Nyquist rate. However, the drawback of the multi-coset approach is that the channel synchronization should be met such that accurate time offsets between sampling channels are required to satisfy a specific sampling pattern. To relax the multichannel synchronization requirement, an asynchronous multirate wideband sensing approach was studied in [179]. The multi-coset sub-Nyquist sampling has been carefully analyzed in [219]. The complexity and the power consumption of these implementations are discussed in detail.

## 3.3 COOPERATIVE SPECTRUM SENSING

Cooperative spectrum sensing has been shown to be an effective method to combat the adverse effects of multipath fading and shadowing by exploiting spatial diversity [74]. Traditionally, the cooperative spectrum sensing methods focused on narrowband sensing [224]. Cooperative spectrum sensing can be classified as either centralized [74, 198, 202] or distributed [76, 115]. In centralized CSS, all the CSUs send their sensing information to a central entity called fusion center (FC). The FC combines the received local sensing information and then takes decision on the occupancy of the PU channel. Distributed cooperative sensing is another approach where instead of reporting them to a FC, the CSUs exchange sensing information with one another. Data fusion is a process of combining local sensing information for hypothesis testing. In general, the sensing results reported to the FC or shared with neighboring users can be combined in two different ways (i) Soft Decision Combining [125, 126, 237] in which the CR users can transmit the entire local sensing samples or the complete local test statistics for soft decision, can utilize the conventional combining techniques such as equal gain combining (EGC) and maximal ratio combining (MRC) and (ii) Hard Decision Combining [151, 238] where CSUs make a local decision and transmit the one bit information for hard combining. The three rules that are applied by hard decision combining are the OR, AND, and $k$ out of $M$ rule.

The performance of cooperative spectrum sensing scheme based on hard combining over Nakagami-$m$ fading channel is studied in [50]. In [144], the performance using energy detection is investigated to improve the sensing performance in channels such as Nakagami fading and log-normal shadowing. Here, the hard decision combining rule is performed at fusion center (FC) to make the final decision on the ON/OFF status of PU. Comparison among data fusion rules investigated for a wide range of average SNR values is also studied in [144]. A similar analysis is done in [143] over Hoyt/Nakagami-$q$ fading channel. The analysis for hard combining over Hoyt and Weibull fading channel is carried out in [142]. Authors in [141] study the CSS using energy detection with soft combining. Here, the performance is studied under several soft data fusion schemes namely, square law selection (SLS), square law combining (SLC) and maximal ratio combining (MRC), that are implemented at the fusion center (FC). The performance has been assessed under AWGN, log-normal shadowing as well as by considering Rayleigh and Rician fading channels.

The performance of energy-detection-based spectrum sensing is greatly affected uncertainty of noise variance which makes the detector unreliable due to "SNR walls". The detection performance of the CSS based on soft combining, where the cooperating CR nodes experience different noise power uncertainties is studied in [82]. Authors here present a detection scheme that is more robust to noise uncertainties than the conventional schemes. In [44], CSS with adaptive thresholds is proposed to improve the detection performance under noise uncertainty. In this algorithm, each SU uses a two-threshold detector for local detection where the threshold at each SU is chosen according to the noise uncertainty at that SU. After each detection, the detection results are fused to give the final decision. In [206], the performance under noise uncertainty is analyzed and a new approach to obtain the SNR wall is proposed. In addition, a suboptimal cooperative sensing algorithm with wavelet denoising is proposed to reduce the impact of noise uncertainty. The SNR wall phenomenon under CSS using AND/OR hard decision with EGC soft decision is analyzed. Authors in [139] develop a sequential probability ratio test for the fuzzy hypotheses testing (FHT) and propose a cooperative sequential spectrum sensing scheme to reduce the effects of noise uncertainty. In this scheme, every cognitive radio computes FHT and subsequently the fusion center combines these fuzzy statistics and decides about the sensing time. The effectiveness of the proposed algorithm is demonstrated using simulations. It is shown that the sample complexity is significantly reduced when compared to the energy detector, sequential crisp hypothesis-testing detector, and fixed sample size FHT detector. The probability of detection for energy detection under $k - \mu$ shadow fading channel is derived in [40]. It is shown that the derived expression for detection probability is a convex function of detection threshold which is utilized to study the latency involved in cooperative spectrum sensing. To reduce the overall system latency, a dynamic spectrum sensing cycle is proposed.

The cooperative spectrum sensing can also be used for wideband sensing. An overview of the challenges and possible solutions for the design of cooperative wideband spectrum sensing in CR networks is presented in [154]. In [155], the spectrum sensing problem is formulated as a class of optimization problems that maximize the aggregated opportunistic throughput of a cognitive radio system under the constraints as interference to the primary users. The problem is mapped into an optimization problem, for which suboptimal solutions are obtained through mathematical transformation under the conditions of practical interest. An expectation maximization (EM) based joint detection and estimation (JDE) scheme for cooperative spectrum sensing in multiuser multiantenna CR network is proposed in [16], where multiple spatially separated SUs cooperate to detect the state of occupancy of a wideband frequency spectrum. The implementation of a compressive-sensing-based cooperative wideband sensing is addressed in [122]. In this technique, the implementation of the conventional architecture that relies on fast Fourier transform (FFT) engine has been adopted and modified to include the multi-coset sampling. As a cooperative wideband sensing, the "frugal sensing" has been proposed in [132] to reduce the bandwidth requirements for the control channel. Authors in [127] assumes a jointly space nature of multiband signals and propose a sub-Nyquist wideband spectrum sensing algorithm that blindly locates the support of occupied channels by recovering the signal support. An efficient cooperative wideband spectrum sensing scheme is proposed that exploits the common signal support shared among the cooperating secondary users. The proposed sub-Nyquist wideband spectrum sensing algorithm is theoretically analyzed and verified using numerical analysis. The performance is also tested on the real-world TV white space signals. It is demonstrated that the proposed scheme performs better than conventional cooperative wideband spectrum sensing scheme and in addition has reduced computational and implementation complexity. Quite a number of studies on cooperative wideband spectrum sensing can also be found in [55, 128, 176, 180, 196].

## 3.4 MACHINE-LEARNING-BASED SPECTRUM SENSING

Spectrum sensing can be modeled as a binary classification problem where secondary users have to classify between presence and absence of the primary users. The machine learning techniques have shown excellent performance in the area of pattern classification. Recently, machine-learning-based spectrum sensing techniques have gained popularity. Authors in [191] proposed cooperative spectrum sensing techniques for cognitive radio network based on unsupervised and supervised machine learning techniques. Unsupervised approaches such as K-means clustering and Gaussian mixture model (GMM) are adopted to develop spectrum sensing algorithms.

For supervised spectrum sensing algorithms support vector machine (SVM) and K-nearest-neighbor (KNN) are used. In [235], a machine-learning-based sensing algorithm is proposed under the scenario where primary user can transmit at multiple power levels. The proposed algorithm combines unsupervised and supervised machine learning approaches for sensing and primary power recognition in multiple primary transmit power (MPTP) scenario. First, the K-means clustering unsupervised learning algorithm is applied to discover the primary user's transmission pattern. Then, the support vector machine is applied to train the CR to determine the status of primary users. The received signal energies are used as the features to train the machine learning algorithm. A similar approach for spectrum sensing using sample covariance matrix of the received signal vector from multiple antennas is proposed in [83]. A machine-learning-based classification is proposed in [119] where a low dimensional probability vector is used as a feature vector instead of $N$ dimensional energy vector. This reduced dimension feature vectors help in reducing the training time of the classifier and also reduce the classification time for testing. A naive Bayes classifier (NBC) based spectrum sensing algorithm is proposed in [192] where spectrum sensing process is formulated into an SNR-related multi-class classification problem.

Deep learning is subset of machine learning area which has outperformed traditional machine learning techniques. Recently, researches have started exploring deep learning algorithms for spectrum sensing. A deep learning based spectrum sensing algorithm is proposed in [70]. Specifically, convolutional long short-term deep neural network (CLDNN) is proposed which is applicable for arbitrary types of primary user signals. The algorithm based on cooperative detection is also proposed to improve the detection performance. In [117, 118], covariance matrix aware convolution neural network (CNN) based deep algorithm for spectrum sensing is proposed where sample covariance matrix is used as an input to CNN. Deep cooperative sensing (DCS) based on convolutional neural network is proposed in [109] where the combining strategy for sensing results from secondary users is learned autonomously with a CNN. A CNN based deep learning approach for spectrum sensing is proposed in [215]. The proposed algorithm takes into account the present and the historical sensing data with which the PU activity pattern is learned to detect the PU activity. Authors in [111] investigate various deep learning algorithms for spectrum sensing applications. They compared classical methods for spectrum sensing and various machine learning algorithms with various deep learning approaches. It is demonstrated using experiments that machine learning algorithms outperform conventional spectrum sensing methods. Also, the three layer CNN offers a better tradeoff between accuracy and computational complexity. In [163], Deep Reinforcement Learning (DRL) based cooperative spectrum sensing algorithm is proposed to decrease the signaling overhead in the network of SUs. Use of generative adversarial network (GAN) for spectrum sensing is discussed in [54]. Two major challenges addressed using GAN are (1) to generate additional synthetic data to improve classification accuracy and (2) to adapt training data to spectrum dynamics. The application of deep learning algorithms to cooperative spectrum sensing is investigated in [89] where the energy efficiency of the distributed cooperative spectrum sensing is investigated by formulating it as a combinatorial optimization problem. Based on this formulation, the use of graph neural network and reinforcement learning are utilized to improve overall energy efficiency.

# Part I

## Narrowband Spectrum Sensing

# 4 Energy-Detection-Based Spectrum Sensing over Generalized Fading Model

In the literature, many studies have been dedicated to the analysis of energy-detection-based spectrum sensing by considering different communication scenarios. In [199], Urkowitz derived the probability of detection ($P_D$) and probability of false alarm ($P_F$) under additive white Gaussian noise channel. He found that decision statistic follows central chi-square distribution when the PU is inactive and when it is active it has non-central chi-square behavior. Kostylev in [105], revisited the problem of ED considering the fading conditions. He derived $P_D$ and $P_F$ under Rayleigh, Rice and Nakagami fading channels. The closed form expressions for the average probability of detection ($\bar{P}_D$) for both single channel and diversity combining scenario under Rayleigh, Rician and Nakagami fading channels are derived in [60]. The performance of ED over wireless channels with composite multipath fading and shadowing is studied in [19]. These effects were modeled by $K$ and $K_G$ channel models. Here, the closed form expressions for $\bar{P}_D$ for no diversity were derived and analysis was then extended to diversity scenario. The performance of energy detector over $\eta-\mu$ fading channel is analyzed in [18]. The authors in [170] discussed the ED performance over $\kappa-\mu$ and $\kappa-\mu$ extreme fading.

The $\eta$-$\lambda$-$\mu$ distribution is a generalized fading model and provides better characterization of the practical channel conditions [149]. In addition to Rayleigh and Nakagami Fading channels, other fading channels occurring in practice can be modeled by adjusting $\eta$, $\lambda$ and $\mu$. Numerical results in [149] show that the $\eta$-$\lambda$-$\mu$ fading model provides better fit to the experimental data than the other models available in the literature. Also, since the shape of the distribution is set by three parameters, it is more flexible. This distribution includes Rayleigh, Rician, Nakagami-$m$, Hoyt, $\eta-\mu$, $\lambda-\mu$, etc distributions as special cases. In spite of the usefulness of this model, no work related to the ED considering this fading is reported in the literature. These factors motivate us to study the performance of the energy-detection-based spectrum sensing by considering the $\eta-\lambda-\mu$ fading. In this chapter, we provide the analysis to ED over $\eta$-$\lambda$-$\mu$ fading channel. We first derive the novel expression for $\bar{P}_D$ for the case of no diversity. The work is then extended to include selection diversity and cooperative spectrum sensing (CSS) scenarios. The $\bar{P}_D$ under composite multipath fading and shadowing is derived. The analysis includes analysis for $K$ and $K_G$ channel model derived in [19] as special cases.

DOI: 10.1201/9781003088554-4

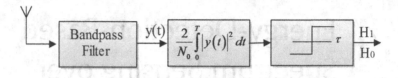

**Figure 4.1** Block diagram of energy detection.

## 4.1 SYSTEM AND CHANNEL MODELS

In this section, the basic idea behind energy detection is discussed. The system model is discussed along with the performance metrics for the energy detection. The performance metrics are also derived without considering fading effects. Finally, the fading model for $\eta - \lambda - \mu$ channel is discussed.

### 4.1.1 ENERGY DETECTION (ED)

The received signal at the secondary user can be represented as [60]

$$y(t) = \begin{cases} n(t); & H_0, \\ h \cdot s(t) + n(t); & H_1, \end{cases} \tag{4.1}$$

where, $s(t)$ is an unknown deterministic signal, $h$ denotes the amplitude of the channel coefficient and $n(t)$ is an additive white Gaussian noise process. The hypotheses $H_0$ and $H_1$ correspond to absence and presence of the primary signal $s(t)$. The basic block diagram of the energy detection is shown in Fig. 4.1. The received signal $y(t)$ is filtered by the bandpass filter (BPF) with bandwidth $W$ Hz which removes the out of band noise power. The output of the filter is then squared, integrated and multiplied by $2/N_0$, which is expressed as $T_m = (2/N_0) \int_0^T |y(t)|^2 dt$ [60], where $N_0$ (W/Hz) is the one sided power spectral density of the AWGN at the receiver and $T$ is the observation interval. The multiplication factor $2/N_0$ is utilized to normalize the noise variance. Under AWGN channel, the decision statistic ($T_m$) follows central chi-square distribution under $H_0$ and non-central chi-square distribution with $N = 2u$ degrees of freedom under $H_1$ [60], where $u = TW$ is the time-bandwidth product. Therefore we have

$$T_m \sim \begin{cases} \chi_N^2; & H_0 \\ \chi_N^2(2\gamma); & H_1, \end{cases} \tag{4.2}$$

where, $\gamma = h^2 E_s / N_0$ is the signal to noise ratio at the input of secondary user, $E_s$ denotes the signal energy.

At the end of the interval $(0, T)$, the detector decides occupancy status of the primary user by comparing the measured energy with the predefined energy threshold

$\tau$. If $T_m \geq \tau$ then the PU is present otherwise it is absent. The $P_F$ and $P_D$ are then obtained as [60]

$$P_F(\tau) = Pr\{T_m > \tau|H_0\} = \frac{\Gamma(\frac{N}{2}, \frac{\tau}{2})}{\Gamma(\frac{N}{2})} \text{ and} \quad (4.3)$$

$$P_D(\tau) = Pr\{T_m > \tau|H_1\} = Q_{\frac{N}{2}}\left(\sqrt{2\gamma}, \sqrt{\tau}\right), \quad (4.4)$$

where, $\Gamma(x) = \int_0^\infty t^{x-1} e^{-t} dt$ is the gamma function [79], $Q_{\frac{N}{2}}(\cdot, \cdot)$ is the generalized Marcum $Q$-function [146], $\Gamma(a,x) = \int_x^\infty t^{a-1} e^{-t} dt$ and $\gamma(a,x) = \int_0^x t^{a-1} e^{-t} dt$ represent upper and lower incomplete gamma functions [4], respectively. The steps involved in deriving Eq.(4.4) is given in APPENDIX B.1. Similar procedure can be followed to derive the expression given in Eq. (4.4).

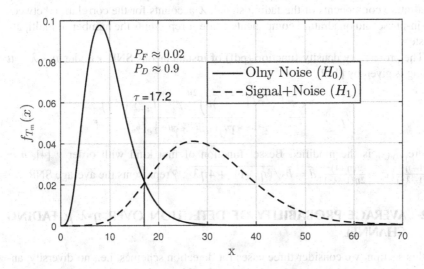

**Figure 4.2** Pictorial understanding of energy detection. The plot is obtained using $N = 10$ and $\gamma = 10\ dB$.

Fig. 4.2 demonstrates the pictorial representation of energy detection. The probability density functions under two hypotheses $H_0$ (only noise) and $H_1$ (signal+noise) are plotted. The plot with a solid line represents the pdf under $H_0$, and the one with the dotted line represents the pdf under $H_1$. One can notice the pdf shift under $H_1$ because of the presence of the PU signal. The plot is obtained for $\gamma = 10\ db$. It should be noted here that with an increase in $\gamma$, i.e., the signal-to-noise ratio, the plot under $H_1$ will further shift to the right side. Similarly, with a decrease in $\gamma$, the plot under $H_1$ will shift to the left. We need to decide the threshold based on the required values of the probability of false alarm and detection. Ideally, we want the probability of false alarm to be zero and the probability of detection to be one. But because of the overlap of the pdfs under $H_0$ and $H_1$, $P_F = 0$ and $P_D = 1$ are impossible to achieve. This is possible only under very high SNR conditions, which is not always possible due to phenomena such as multipath fading and shadowing in the wireless environment. Hence, one needs to allow a certain value of false alarm to decide the threshold, and the probability of detection is computed using that threshold. One such value is taken as an example in Fig. 4.2. The value of allowed false alarm is taken as $P_F \approx 0.02$ and based on that the value of threshold is computed to be $\tau = 17.2$. For the selected threshold, the probability of false alarm is computed to be $P_D \approx 0.9$.

### 4.1.2 $\eta$-$\lambda$-$\mu$ FADING MODEL

In [149], a new general fading model for mobile communication is proposed. This model combines the properties of the $\lambda$-$\mu$ and $\eta$-$\mu$ distribution models. As already discussed, this is a flexible model since it is defined in terms of three parameters $\eta$, $\lambda$ and $\mu$. Here, the parameter $\eta$ accounts for unequal powers of the in-phase and quadrature components of the fading signal, $\lambda$ accounts for the correlation between the in-phase and quadrature components while $\mu$ represents the number of multipath clusters.

The probability density function (pdf) of instantaneous SNR $\gamma$ under $\eta - \lambda - \mu$ fading is given by [149]

$$f_\gamma(\gamma) = \frac{\sqrt{\pi}\left(\sqrt{\eta(1-\lambda^2)}\tilde{b}\right)^{2\mu} \gamma^{\mu-\frac{1}{2}} I_{\mu-\frac{1}{2}}\left(\tilde{d}\frac{\gamma}{\bar{\gamma}}\right)}{2^{-\mu-\frac{1}{2}}\Gamma(\mu)\tilde{d}^{\mu-\frac{1}{2}}\bar{\gamma}^{\mu+\frac{1}{2}}e^{\tilde{c}\frac{\gamma}{\bar{\gamma}}}}, \tag{4.5}$$

where, $I_\nu(\cdot)$ is the modified Bessel function of first kind with order $\nu$ [4], $\tilde{b} = \frac{\mu(1+\eta)}{2\eta(1-\lambda^2)}$, $\tilde{c} = \frac{\mu(\eta+1)^2}{2\eta(1-\lambda^2)}$, $\tilde{d} = \tilde{b}\sqrt{(\eta-1)^2 + 4\eta\lambda^2}$, $\bar{\gamma}$ represents the average SNR.

## 4.2 AVERAGE PROBABILITY OF DETECTION OVER $\eta$-$\lambda$-$\mu$ FADING CHANNEL

In this section, we consider three cases for detection schemes, i.e., no diversity, antenna diversity, and cooperative detection. In this work, we aim to show that the performance can be improved by using antenna diversity and using cooperative detection. Hence, we derive $\bar{P}_D$ considering square law selection (SLS) only. One can also derive $\bar{P}_D$ considering other diversity techniques such as square law combining (SLC), equal gain combining (EGC), and maximum ratio combining (MRC). Similarly, we derive $\bar{P}_D$ for cooperative detection using OR hard combining only, which can be extended to other hard combining techniques such as AND, $k$ out of $M$, and majority combining techniques.

### 4.2.1 NO DIVERSITY

In this section, the probability of detection is derived under no diversity case considering $\eta - \lambda - \mu$ fading. The Eq. (4.3) and Eq. (4.4) give the probability of false

# Energy-Detection-Based Spectrum Sensing over Generalized Fading Model 71

alarm ($P_F$) and probability of detection ($P_D$) under AWGN channel, respectively. The probability of false alarm remains same under fading conditions since it is independent of SNR and $H_0$ corresponds to only noise signal. The probability of detection is greatly affected by the fading phenomenon. We need to carefully derive the probability of detection considering fading in order to understand the detector performance. The probability of detection under fading channel is obtained by averaging $P_D$ under AWGN channel in Eq. (4.5) over probability density function of SNR under fading channel, i.e.,

$$\bar{P}_D = \int_0^\infty Q_{\frac{N}{2}}(\sqrt{2\gamma}, \sqrt{\tau}) f_\gamma(\gamma) d\gamma. \tag{4.6}$$

The pdf of SNR under $\eta - \lambda - \mu$ fading channel is given in Eq. (4.5). Substituting Eq. (4.5) into Eq. (4.6), we get

$$\bar{P}_D = \int_0^\infty \frac{Q_{\frac{N}{2}}(\sqrt{2\gamma}, \sqrt{\tau})\sqrt{\pi}\left(\sqrt{\eta(1-\lambda^2)}\tilde{b}\right)^{2\mu} I_{\mu-\frac{1}{2}}\left(\tilde{d}\frac{\gamma}{\tilde{\gamma}}\right)}{2^{-\mu-\frac{1}{2}}\Gamma(\mu)\tilde{d}^{\mu-\frac{1}{2}}\tilde{\gamma}^{\mu+\frac{1}{2}} e^{\tilde{c}\frac{\gamma}{\tilde{\gamma}}} \gamma^{-\mu+\frac{1}{2}}} d\gamma. \tag{4.7}$$

The generalized Marcum Q function in Eq. (4.7) can be expressed in series form as

$$Q_{\frac{N}{2}}(\sqrt{2\gamma}, \sqrt{\tau}) = e^{-\gamma} \sum_{l=0}^\infty \frac{\gamma^l \Gamma(l+u, \frac{\tau}{2})}{\Gamma(l+1)\Gamma(l+u)}. \tag{4.8}$$

The series form of modified Bessel function of first kind used in Eq. (4.7) can be given as [79],

$$I_v(x) = \sum_{j=0}^\infty \frac{x^{2j+v}}{j! 2^{2j+v} \Gamma(j+v+1)}, \tag{4.9}$$

Using Eq. (4.8) and Eq. (4.9) in Eq. (4.7), we can write $\bar{P}_D$ as

$$\bar{P}_D = \sum_{l=0}^\infty \sum_{j=0}^\infty \frac{\sqrt{\pi}(\sqrt{\eta(1-\lambda^2)}\tilde{b})^{2\mu} \Gamma(l+u, \frac{\tau}{2})(\tilde{d}^{2j})}{2^{2j-1}\Gamma(\mu)\tilde{\gamma}^{2j+2\mu} l! j! \Gamma(l+\frac{N}{2})\Gamma(j+\mu+\frac{1}{2})} \int_0^\infty \gamma^{l+2\mu+2j-1} e^{-\gamma(1+\frac{\tilde{c}}{\tilde{\gamma}})} d\gamma. \tag{4.10}$$

Now consider the integral in Eq. (4.10) which is represented by

$$I = \int_0^\infty \gamma^{l+2\mu+2j-1} e^{-\gamma(1+\frac{\tilde{c}}{\tilde{\gamma}})} d\gamma. \tag{4.11}$$

Applying the change of variable as $x = \gamma(1+\frac{\tilde{c}}{\tilde{\gamma}})$, this integral reduces to

$$I = (1+\frac{\tilde{c}}{\tilde{\gamma}})^{-(l+2\mu+2j)} \int_0^\infty x^{l+2\mu+2j-1} e^{-x} dx,$$

$$= (1+\frac{\tilde{c}}{\tilde{\gamma}})^{-(l+2\mu+2j)} \Gamma(l+2j+2\mu). \tag{4.12}$$

Substituting back Eq. (4.12) into Eq. (4.11) the series representation for $\bar{P}_D$ can be written as

$$\bar{P}_D = \sum_{l=0}^{\infty}\sum_{j=0}^{\infty} \frac{\bar{\gamma}^l\sqrt{\pi}(\sqrt{\eta(1-\lambda^2)}\tilde{b})^{2\mu}\Gamma(l+u,\frac{\tau}{2})\tilde{d}^{2j}\Gamma(l+2j+2\mu)}{2^{2j-1}\Gamma(\mu)l!j!\Gamma(l+\frac{N}{2})\Gamma(j+\mu+\frac{1}{2})(\bar{\gamma}+\tilde{c})^{l+2\mu+2j}}. \quad (4.13)$$

The series in Eq. (4.13) converges for $l \to \infty$ and $j \to \infty$. However, we observe that the series nearly converges when the sum is considered over a relatively small values of $l$ and $j$ and can be computed easily using a mathematical software (e.g. MATHEMATICA [213]). For example, with $\eta = 0.4$, $\lambda = 0.5$, $\mu = 1$, $N = 10$, $\bar{\gamma} = 10$ and $P_F = 0.1$, the series in Eq. (4.13) converges upto four decimal points for $l = 79$ and $j = 11$. Hence, $\bar{P}_D$ for these parameters can be obtained by taking the finite sum as

$$\bar{P}_D = \sum_{l=0}^{79}\sum_{j=0}^{11} \frac{\bar{\gamma}^l\sqrt{\pi}(\sqrt{\eta(1-\lambda^2)}\tilde{b})^{2\mu}\Gamma(l+u,\frac{\tau}{2})\tilde{d}^{2j}\Gamma(l+2j+2\mu)}{2^{2j-1}\Gamma(\mu)l!j!\Gamma(l+\frac{N}{2})\Gamma(j+\mu+\frac{1}{2})(\bar{\gamma}+\tilde{c})^{l+2\mu+2j}}. \quad (4.14)$$

### 4.2.2 SQUARE LAW SELECTION (SLS) DIVERSITY

In wireless communication, antenna diversity is used to tackle the adverse effects of fading. In antenna diversity, the number of spatially separated antennas, also known as diversity branches, is used at the receiver. The spacing between the antennas is kept such that all the antennas receive independent signals. Generally, the spacing between the antennas is selected as $\frac{\lambda}{2}$. Since all the diversity branches receive independent observations, the probability that all the branches experience deep fades at the same time reduces with an increase in the number of diversity branches. Hence, if one or more diversity branches are in a deep fade, we can use signal received at the other diversity branches that do not experience deep fading for signal detection. The same diversity concept is used in spectrum sensing techniques to improve detection performance under fading. There are multiple ways in which signals from multiple diversity branches can be combined at the secondary users such as SLS, SLC, EGC, and MRC. In this section, the use of the square law selection diversity combining technique is discussed, and the probability of detection is derived considering $\eta - \lambda - \mu$ fading channel.

The SLS diversity technique is demonstrated using a block diagram in Fig. 4.3. There are $P$ number diversity branches and the decision statistic, i.e., received energy, is computed at all the diversity branches. As shown in Fig. 4.3, the branch with maximum decision statistic is chosen [145], i.e., $T_{SLS} = \max(T_1, T_2, \ldots, T_P)$, where $T_{SLS}$ is the selected diversity branch. The probability of detection ($P_D^{SLS}$) under AWGN channel with SLS diversity scheme is derived in [60] and is given by

$$P_D^{SLS} = 1 - \prod_{i=1}^{P}[1 - P_{D_i}],$$

$$= 1 - \prod_{i=1}^{P}\left[1 - Q_{\frac{N}{2}}(\sqrt{2\gamma_i}, \sqrt{\tau})\right], \quad (4.15)$$

where, $\gamma_i$ is the signal to noise radio at the $i^{th}$ diversity branch and $P_{D_i}$ is the probability of detection computed using decision static at the $i^{th}$ diversity branch.

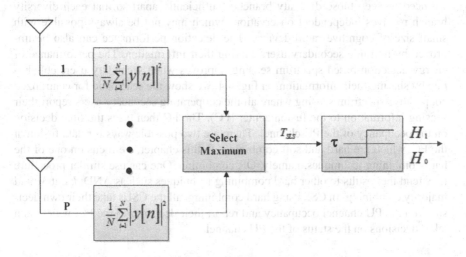

**Figure 4.3** Block diagram of SLS diversity technique.

Based on this, $\bar{P}_D^{SLS}$ in the case of $\eta$-$\lambda$-$\mu$ can be obtained by averaging $P_D^{SLS}$ over $P$ independent branches, i.e.,

$$\bar{P}_D^{SLS} = 1 - \int_0^\infty \cdots \int_0^\infty \prod_{i=0}^{P} \left[1 - Q_{\frac{N}{2}}(\sqrt{2\gamma_i}, \sqrt{\tau})\right] f_{\gamma_i}(\gamma_i) d\gamma_i. \quad (4.16)$$

Since $\int_0^\infty f_{\gamma_i}(\gamma_i) d\gamma_i \triangleq 1$, the average probability of detection with SLS diversity is obtained as

$$\bar{P}_D^{SLS} = 1 - \prod_{i=1}^{P} \left[1 - \int_0^\infty Q_{\frac{N}{2}}(\sqrt{2\gamma_i}, \sqrt{\tau}) f_{\gamma_i}(\gamma_i) d\gamma_i\right]. \quad (4.17)$$

Note that the integral that needs to be evaluated in Eq. (4.17) is same as that of no diversity case. Therefore, following the same steps, $\bar{P}_D^{SLS}$ can obtained as

$$\bar{P}_D^{SLS} = 1 - \prod_{i=1}^{P} \left[1 - \sum_{l=0}^{\infty} \sum_{j=0}^{\infty} \frac{\bar{\gamma}_i^l \sqrt{\pi} (\sqrt{\eta(1-\lambda^2)} \tilde{b})^{2\mu} \Gamma(l+\frac{N}{2},\frac{\tau}{2}) \tilde{d}^{2j} \Gamma(l+2j+2\mu)}{2^{2j-1} \Gamma(\mu) l! j! \Gamma(l+\frac{N}{2}) \Gamma(j+\mu+\frac{1}{2}) (\tilde{\gamma}+\tilde{c})^{l+2\mu+2j}}\right]. \quad (4.18)$$

The probability of false alarm in case of SLS remains same as that of AWGN channel since it is independent of SNR and is given by [60, Eq. (14)]

$$P_F^{SLS} = 1 - [1 - P_F]^P = 1 - \left[1 - \frac{\Gamma(\frac{N}{2}, \frac{\tau}{2})}{\Gamma(\frac{N}{2})}\right]^P. \quad (4.19)$$

### 4.2.3 COOPERATIVE SPECTRUM SENSING

Using diversity to improve performance requires multiple diversity branches. Also we need to keep those diversity branches sufficiently apart so that each diversity branch receives independent observations which may not be always possible with small size of cognitive radio devices. The detection performance can also be improved by multiple secondary users sharing their information. The performance of energy-detection-based spectrum sensing improves when secondary users collaborate by sharing their information. In Fig. 4.4, we show the schematic for centralized cooperative spectrum sensing where all the cooperating secondary users report their sensing information to the fusion center (FC). The FC then takes the final decision on the occupancy of the PU channel. There are two possible ways for data fusion at the FC, which are hard and soft combining. In this chapter, we focus on one of the hard combining techniques, namely, OR combining. One can use similar procedure to extend the results to other hard combining techniques such as AND, $k$ out $M$ and majority combining. In CSS using hard combining, all the CSUs take their own decisions on the PU channel occupancy and report their decisions to the FC, which then takes decisions on the status of the PU channel.

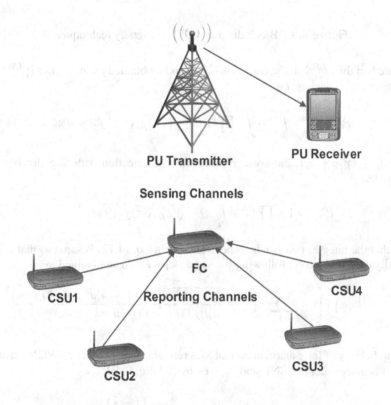

**Figure 4.4** Centralized cooperative spectrum sensing.

In the OR combining scheme, if any one CSU reports the PU channel as occupied, the FC declares the PU channel as occupied. The block diagram for CSS using OR combining is shown in Fig. 4.5 with three CSUs where CSU 1 has reported channel as occupied, and the other two CSUs have reported channel as free. The FC has declared PU channel as occupied since one of the CSUs has reported the channel as occupied. In this scenario, the average probability of detection and false alarm considering OR hard combining at the FC with $M$ independent CSUs are given by [75]

$$\bar{Q}_D \triangleq 1 - (1 - \bar{P}_D)^M \text{ and } Q_F \triangleq 1 - (1 - P_F)^M. \qquad (4.20)$$

Here, we assume that all the CSUs experience the same average signal to noise ratios and the same noise distributions.

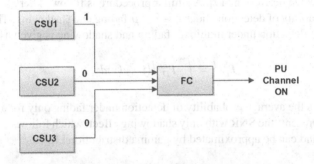

**Figure 4.5** Block diagram for CSS using OR combining.

The average probability of detection with $M$ cooperating secondary users is obtained by substituting Eq. (4.13) into $\bar{Q}_D$ in Eq. (4.20) as

$$\bar{Q}_D = 1 - \left[ 1 - \sum_{l=0}^{\infty} \sum_{j=0}^{\infty} \frac{\bar{\gamma}^l \sqrt{\pi}(\sqrt{\eta(1-\lambda^2)}\bar{b})^{2\mu} \Gamma(l+u, \frac{\tau}{2}) \bar{d}^{2j} \Gamma(l+2j+2\mu)}{2^{2j-1}\Gamma(\mu) l! j! \Gamma(l+\frac{N}{2}) \Gamma(j+\mu+\frac{1}{2}) (\bar{\gamma}+\bar{c})^{l+2\mu+2j}} \right]^M. \qquad (4.21)$$

As already stated $P_F$ remains same as under AWGN channel and hence $Q_F$ is given by

$$Q_F = 1 - (1 - P_F)^M,$$

$$= 1 - \left( 1 - \frac{\Gamma(\frac{N}{2}, \frac{\tau}{2})}{\Gamma(\frac{N}{2})} \right)^M. \qquad (4.22)$$

## 4.3 AVERAGE PROBABILITY OF DETECTION OVER CHANNELS WITH $\eta - \lambda - \mu$ FADING AND SHADOWING

Apart from the multipath fading, the signal received at the SUs also undergoes shadowing. A lognormal distribution [178] typically models the shadowing process. Therefore, some practical communication channels can be modeled as multipath fading superimposed on lognormal shadowing. Due to the difficulty of analyzing spectrum sensing techniques over composite fading models, the shadowing effect is usually neglected in the literature. In this section, we derive the expression for the average probability of detection where the signal undergoes shadowing in addition to $\eta - \lambda - \mu$ fading. In [3], Gamma distribution is used to approximate lognormal distribution to derive K and generalized K ($K_G$) distribution as a composite multipath fading and shadowing. The $K$ channel model corresponds to the mixture of Rayleigh distribution and gamma distribution, while $K_G$ represents a mixture of Nakagami and Gamma distributions. The average probabilities of detection under $K$ and $K_G$ channel models have been derived in [19]. A similar procedure is followed here to derive the average probability of detection under $\eta - \lambda - \mu$ fading and shadowing. The average probability of detection under multipath fading and shadowing is given as

$$\bar{P}_D^{Shd} = \int_0^\infty \bar{P}_D^{Fad}(y) f_Y(y) dy, \qquad (4.23)$$

where, $\bar{P}_D^{Fad}$ is the average probability of detection under fading only for a specific $Y$ value that represents the SNR with only shadowing effect, which follows a lognormal distribution and can be approximated by gamma distribution [3].

$$f_Y(y) = \frac{y^{k-1} e^{-\frac{y}{\Omega}}}{\Gamma(k)\Omega^k}, \quad y \geq 0, \qquad (4.24)$$

where, $k$ is the shaping parameter and $\Omega$ represents the scale parameter which is also the mean signal power and $\bar{P}_D^{Fad}(y)$ is obtained by replacing every $\bar{\gamma}$ in Eq. (4.13) with $y$. Using this $\bar{P}_D^{Fad}(y)$ in Eq. (4.23) and averaging over pdf in Eq. (4.24), we get $\bar{P}_D^{Shd}$ as

$$\bar{P}_D^{Shd} = \sum_{l=0}^\infty \sum_{j=0}^\infty \frac{\sqrt{\pi}(\sqrt{\eta(1-\lambda^2)}\tilde{b})^{2\mu} \Gamma(l+u,\frac{\tau}{2}) \tilde{d}^{2j}}{2^{2j-1}\Gamma(\mu) l! j! \Gamma(l+\frac{N}{2}) \Gamma(j+\mu+\frac{1}{2})}$$
$$\times \frac{\Gamma(l+2j+2\mu)}{\Gamma(k)\Omega^k} \int_0^\infty y^{l+k-1}(y+\tilde{c})^{-(l+2\mu+2j)} e^{-\frac{y}{\Omega}} dy. \qquad (4.25)$$

The integral in Eq. (4.25) can be written as

$$I = \tilde{c}^{-(l+2\mu+2j)} \int_0^\infty y^{(l+k)-1}\left(1+\frac{1}{\tilde{c}}y\right)^{-(l+2\mu+2j)} e^{-\frac{y}{\Omega}} dy,$$
$$= \Gamma(l+k)\tilde{c}^{k-2\mu-2j} U(l+k; k-2\mu-2j+1; \frac{\tilde{c}}{\Omega}), \qquad (4.26)$$

where, $U(;;)$ is the confluent hypergeometric function of the second kind defined as [79, Eq. (3.383.5)]:

$$\int_0^\infty e^{-px}x^{q-1}(1+ax)^{-v}dx = \frac{\Gamma(q)}{a^q}U(q;q+1-v;\frac{p}{a}), \quad (4.27)$$

with $Re\{q\} > 0$, $Re\{p\} > 0$, $Re\{a\} > 0$ and $v$ is a complex value with $Re(\cdot)$ representing the real operator which gives real part of the complex number.

Substituting Eq. (4.26) into Eq. (4.25), the average detection probability under multipath fading and shadowing can be written as

$$\bar{P}_D^{Shd} = \sum_{l=0}^{\infty} \sum_{j=0}^{\infty} \frac{\sqrt{\pi}(\sqrt{\eta(1-\lambda^2)}\tilde{b})^{2\mu}\Gamma(l+u,\frac{\tau}{2})\tilde{d}^{2j}}{2^{2j-1}\Gamma(\mu)l!j!\Gamma(l+\frac{N}{2})\Gamma(j+\mu+\frac{1}{2})}$$
$$\times \frac{\Gamma(l+2j+2\mu)\Gamma(l+k)U(l+k;k-2\mu-2j+1;\frac{\tilde{c}}{\Omega})}{\Gamma(k)\Omega^k \tilde{c}^{(2\mu+2j-k)}}. \quad (4.28)$$

It may be of interest to note that the detection performance can be improved by using diversity as well as the cooperative detection. Once $\bar{P}_D^{Shd}$ is obtained, the average probability of detection with fading and shadowing under SLS diversity can be obtained by averaging $\bar{P}_D^{SLS}$ in Eq. (4.18) over gamma distribution in Eq. (4.24) after replacing each $\bar{\gamma}_i$ by $y_i$ in Eq. (4.18) and each $y$ by $y_i$ in Eq. (4.24). The average probability of detection under cooperative detection can be obtained by substituting $\bar{P}_D^{Shd}$ from Eq. (4.28) into Eq. (4.20).

## 4.4 RESULTS AND DISCUSSION

In this section, we carry out the analysis to test the performance of energy detection under $\eta$-$\lambda$-$\mu$ fading channel for several cases of interest. The performance is studied using $\bar{\gamma}$ Vs. $\bar{P}_D$ curve, i.e., average SNR Vs. the average probability of detection and complementary receiver operating characteristic (ROC) curves, i.e., $P_F$ Vs. $P_M = 1 - P_D$. The performance improvement using diversity and cooperative spectrum sensing is also demonstrated using ROC plots. Besides, $\eta$-$\lambda$-$\mu$ fading channel provides analysis for other fading channels as special cases.

Fig. 4.6 displays $\bar{\gamma}$ Vs. $\bar{P}_D$ plot for no diversity case under $\eta$-$\lambda$-$\mu$ fading channel for $P_F = 0.1$, $N = 4$, $\eta = 0.4$, $\lambda = 0.5$ and $\mu = 1$. The value of average SNR ($\bar{\gamma}$) is varied from 0 $dB$ to 20 $dB$. As expected, the detection performance improves with increase in SNR. One can note that for very high value of $\bar{\gamma}$ the probability of detection approaches 1. Fig. 4.7 shows the complementary ROC curve, i.e., $P_F$ Vs. $P_M$, using $\bar{\gamma} = 5$ $dB$ and keeping the other parameters same as that of $\bar{\gamma}$ versus $\bar{P}_D$ shown in Fig. 4.6.

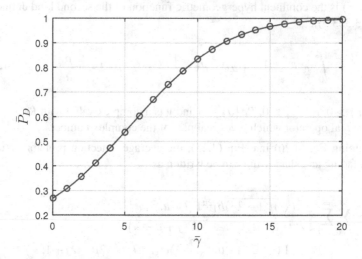

**Figure 4.6** $\bar{\gamma}$ vs $\bar{P}_D$ for no diversity considering for $P_F = 0.1$, $N = 4$, $\eta = 0.4$, $\lambda = 0.5$, $\mu = 1$.

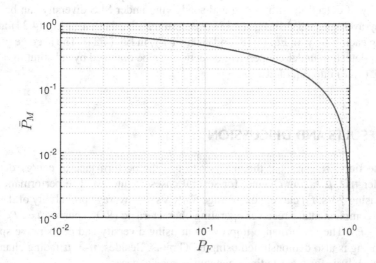

**Figure 4.7** $P_F$ vs $\bar{P}_M$ for no diversity case considering $\bar{\gamma} = 5\ dB$, $N = 4$, $\eta = 0.4$, $\lambda = 0.5$, $\mu = 1$.

The analysis for $\eta$-$\lambda$-$\mu$ fading channel is general since the other fading channels are special cases of $\eta$-$\lambda$-$\mu$ fading distribution [149]. It should be noted that there are other ways to relate various distributions with each other except those described in this section. The Nakagami-$m$ distribution can be obtained by setting $\eta \to 1$, $\lambda \to 0$

and $\mu = 0.5m$. The Rayleigh, One-Sided Gaussian and Nakagami-$q$ (Hoyt) distribution can be obtained from Nakagami-$m$ distribution by setting $m = 1$, $m = 0.5$ and $m = (1+q^2)^2/2(1+2q^4)$ respectively. The $\eta - \mu$ distribution can be obtained by setting $\lambda \rightarrow 0$ and $\mu = \mu'$, where $\mu'$ corresponds to the simpler model. Similarly, the $\lambda - \mu$ distribution can be obtained by setting $\eta \rightarrow 1$ and $\mu = \mu'$. In Fig. 4.8 we display the complementary ROC curves for different fading channels obtained from $\eta$-$\lambda$-$\mu$ fading channel as special cases discussed above. For this analysis we used $SNR = 10\ dB$ and $N = 10$.

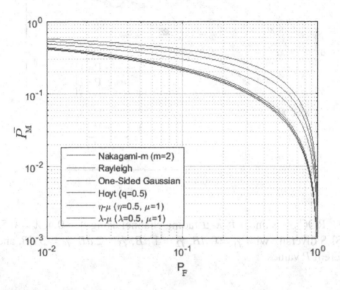

**Figure 4.8** $P_F$ vs $\bar{P}_M$ for other fading channels as a special case of $\eta$-$\lambda$-$\mu$ fading distribution with $\bar{\gamma} = 10\ dB$ and $N = 10$.

**Table 4.1**
**Effect of Increasing Number of Diversity Branches on $(\bar{P}_D)$ on the Performance**

| Values of $\bar{P}_D$ for different values $P$ with fixed $P_F = 0.1$ | | | | | |
|---|---|---|---|---|---|
| P | 1 | 2 | 3 | 4 | 5 |
| $\bar{P}_D$ | 0.25 | 0.33 | 0.45 | 0.52 | 0.63 |

Fig. 4.9 shows plots of $P_F$ Vs. $\bar{P}_D^{SLS}$ by varying the number of diversity branches $P$ from 1 to 5 where $P = 1$ corresponds to no diversity. For this analysis, we used $\eta = 0.4$, $\lambda = 0.5$, $\mu = 1$, $N = 4$ and the average SNRs are set to $\bar{\gamma}_1 = 0\ dB$, $\bar{\gamma}_2 = 1\ dB$, $\bar{\gamma}_3 = 2\ dB$, $\bar{\gamma}_4 = 3\ dB$, and $\bar{\gamma}_5 = 4\ dB$. Here, we see that the detection performance

improves with the increase in the number diversity branches. To illustrate the effect numerically, the values of $\bar{P}_D$ for fixed value of $P_F$, i.e., $P_F = 0.1$, is tabulated in TABLE 4.1 for different values of $P$. From the table we can see the performance improvement as $P$ increases. For example, for $P_F = 0.1$, the value of $\bar{P}_D$ with $P = 5$ is 152 % higher than that with $P = 1$.

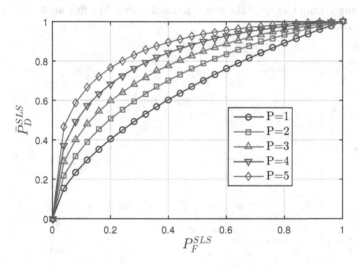

**Figure 4.9** ROC plots under $\eta$-$\lambda$-$\mu$ fading channel using $\eta = 0.4$, $\lambda = 0.5$, $\mu = 1$, $N = 4$ for SLS diversity with $\bar{\gamma}_1 = 0\ dB$, $\bar{\gamma}_2 = 1\ dB$, $\bar{\gamma}_3 = 2\ dB$, $\bar{\gamma}_4 = 3\ dB$, and $\bar{\gamma}_5 = 4\ dB$ for different $P$ values.

**Table 4.2**
**Effect of Increasing Number of Cooperating Secondary Users on $(\bar{P}_D)$ on the Performance**

| | Values of $\bar{P}_D$ for different values $M$ with fixed $P_F = 0.1$ | | | |
|---|---|---|---|---|
| $M$ | 1 | 2 | 4 | 8 |
| $\bar{P}_D$ | 0.27 | 0.31 | 0.4 | 0.48 |

Fig. 4.10 demonstrates the complementary ROC curve for energy detector under $\eta$, $\lambda$ and $\mu$ fading channel considering up to eight cooperating secondary users. The analysis is performed using the same parameters setting as that of the SLS diversity scenario, and the $\bar{\gamma}$ is set to $0\ dB$. As expected, the detection performance improves with the increase in the number of cooperating secondary users. To illustrate the effect numerically, the values of $\bar{P}_D$ for fixed value of $P_F$, i.e., $P_F = 0.1$, is tabulated in TABLE 4.2 for different values of $M$. From the table, we can see the performance

improvement as $M$ increases. For example, for $P_F = 0.1$, the value of $\bar{P}_D$ with $M = 8$ is 77.77 % higher than that with $M = 1$. We can see that the performance improvement with cooperative spectrum sensing without using diversity at the cooperating secondary users. The issue of using multiple spatially separated diversity branches that we face in antenna diversity scheme can be resolved using CSS. We have shown here use of OR hard combining technique only but one can make use of other combining techniques used in the CSS. The use of soft combining can give better performance compared to hard combining. In soft combining actual decision statics are sent by the CSUs to the fusion center which then takes the final decision based on the received observations. However, the control channel bandwidth requirements are very high in soft combining and because of that the use of hard combining is preferred. One can make use of antenna diversity and cooperative spectrum sensing to further improve the detection performance.

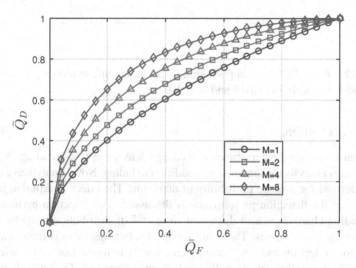

**Figure 4.10** ROC plots under $\eta$-$\lambda$-$\mu$ fading channel using $\eta = 0.4$, $\lambda = 0.5$, $\mu = 1$, $N = 4$ for cooperative spectrum sensing with $\bar{\gamma} = 0\ dB$ and $M$ cooperative secondary users.

Fig. 4.11 shows the plots for $\bar{P}_D^{Shd}$ versus $P_F$ for channel that undergoes shadowing in addition to $\eta - \lambda - \mu$ fading. The plots are obtained by choosing different values for the shadowing parameter $k$, i.e., by choosing $k = 0.5$ and $k = 1$. The analysis includes the analysis for $K$ and $K_G$ channels as special cases. The analysis for $K$ channel model can be obtained by setting $\eta \to 1$, $\lambda \to 0$, $m = 1$ and $\mu = 0.5m$. Similarly, by setting $\eta \to 1$, $\lambda \to 0$, $m = 2$ and $\mu = 0.5m$, the analysis for $K_G$ channel model can be deduced. The results for $K$ and $K_G$ channel models match the results in [19]. The plot also shows results for shadowing in addition to fading with fading parameters $\eta = 0.4$, $\lambda = 0.5$ and $\mu = 1$ for $\bar{\gamma} = 10\ dB$. As already discussed, the detection performance can be improved by using the diversity schemes and cooperative detection.

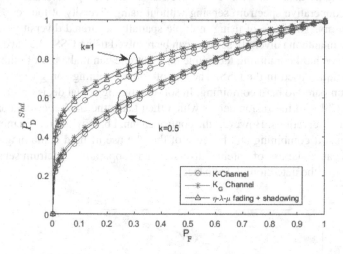

**Figure 4.11** $P_F$ vs $\bar{P}_D^{Shd}$ for composite multipath fading and shadowing channel for $k = 0.5$ and $k = 1$ with $\bar{\gamma} = 10\ dB$ and $N = 10$.

## 4.5 CONCLUSION

This chapter illustrates the performance of energy detector in $\eta$-$\lambda$-$\mu$ fading channel. It also includes analysis for shadowing in addition to fading. Novel analytical expressions are derived for average probability of detection. The effect of different fading parameters on the detection performance is discussed. The detection performance of other fading channels is also discussed since $\eta$-$\lambda$-$\mu$ distribution provides those distributions as special cases. The result for no diversity scenario is then extended to square low selection and cooperative detection. It is found that under selection diversity and cooperative detection, the performance improves. The analysis is then extended to the case when there exists shadowing in addition to $\eta$-$\lambda$-$\mu$ fading.

# 5 Generalized Energy Detector in the Presence of Noise Uncertainty and Fading

In Chapter 4, we have discussed conventional energy detector (CED) based spectrum sensing. In this chapter, we discuss generalized energy detector (GED), which is obtained replacing the squaring operation of the received signal amplitude in CED by an arbitrary positive power $p$. As in CED, here also, the decision on the occupancy of a channel is made based on the predefined threshold. The noise variance can determine a proper value of threshold at the SU, and hence, it plays an important role in determining the detector's performance. In this chapter, we consider the noise uncertainty (NU) in determining the threshold. One has to know the true noise variance to determine the value of this threshold. If known exactly, it is possible to sense the occupancy of PU even at very low SNR if the sensing time is made sufficiently large [101], i.e., a large number of samples ($N$) are used in sensing. In Chapter 4, when we discuss ED, it is assumed that the noise variance is known and remains unchanged. In practice, the noise variance at the SU input varies with time and location, so it is not possible to find its exact value. Due to this, there is an unpredictability about the actual variance of the noise, known as noise uncertainty. The effect of worst-case noise uncertainty is discussed in [173, 189]. In [189], a phenomenon called SNR wall is studied for the CED based sensing method, which says that if the noise variance is not known precisely and is confined to an interval, it is not possible to achieve targeted detection performance when the SNR falls below certain value regardless of sensing time. This makes ED an inefficient sensing method.

The effect of uniformly distributed noise uncertainty on the performance of CED is studied in [82, 232]. The authors in [232] derive the expression for the SNR wall for CED, assuming a uniform distribution for noise uncertainty. The detection performance of the cooperative spectrum sensing (CSS) is studied in [82]. In [130], an asymptotic analysis of noise power estimation is performed for CED. The condition for the existence of the SNR wall is obtained, and the effect of noise power estimation on the performance of CED is studied. In [97, 98], the performance of GED has studied under noise uncertainty wherein the authors in [97] show that under the worst case of noise uncertainty the SNR wall is independent of the value of $p$. It is also shown that under the assumption of uniform distribution of noise uncertainty, the CED represents the optimum ED. The SNR wall expression is obtained in [98], where the noise uncertainty is once again chosen as uniformly distributed.

DOI: 10.1201/9781003088554-5

It is also shown that the SNR wall does not depend on the value of $p$. The study of the detection performance is then extended to noise uncertainty having log normal distribution, and the SNR wall for the same is calculated numerically.

Authors in [97, 98] derive the SNR wall for GED under noise uncertainty considering no diversity. The SNR wall for CSS with CED assuming the same SNRs and the noise uncertainties for all the cooperating secondary users is discussed in [206, 226]. However, in practice, the SNR varies with time and location. The noise uncertainty depends on calibration error, variations in thermal noise, and changes in low nose amplifier gain. Hence, the assumption of the same SNR and noise uncertainty at all the SUs is not valid in practice. The scenario in which different CSUs have the varying noise uncertainties is studied in [44, 82], but the authors do not discuss the SNR wall. We notice that the existing analysis is only limited to no diversity or CSS case. Also, the available analysis does not consider noise uncertainty with fading. We know that diversity is suited to combat the adverse effects of multi-path fading and shadowing [18, 19, 59, 60, 145, 170]. Two new diversity schemes namely $p$−law combining (pLC) and $p$−law selection (pLS) are proposed in [24], in which GED is used at each of the diversity branches. Both pLC and pLS diversity schemes are non-coherent, combining techniques. Here, they consider diversity and fading without considering noise uncertainty. The pLC and pLS diversity schemes are an extension of square law combining (SLC) and square law selection diversity, respectively. This motivates us to analyze the SNR wall for GED under diversity and CSS considering noise uncertainty and fading. For CSS, we consider both hard as well as soft combining. For hard combining, we consider all three possible cases, i.e., OR, AND, and $k$ out of $M$ combining rule. We first derive the SNR wall considering the AWGN channel and then extend it to the channel with Nakagami fading. It is a generalized fading model and includes Rayleigh, Rice, and Hoyt fading models as its special cases and can be used to model propagation in urban and sub-urban areas by setting the Nakagami parameter $m$. In Chapter 4, we analyzed ED based spectrum sensing under $\eta - \lambda - \mu$ fading channel where we did not consider the noise uncertainty. Although the $\eta - \lambda - \mu$ fading model is more general, theoretical analysis considering this model is mathematically too involved, and hence we give theoretical analysis considering Nakagami fading only. To the best of our knowledge, researchers have not worked on GED under diversity and CSS, considering noise uncertainty and fading. As the first study on GED with diversity in the presence of noise uncertainty and fading, the analysis here is limited to pLC and pLS diversity only. We also study the effect of $p$ on detection performance. Finally, the effects on noise uncertainty and fading on the detection performance is studied.

## 5.1 SYSTEM MODEL

In cognitive radio, the signal received at the SU can be written as

$$y(n) = \begin{cases} w(n); & H_0, \\ h(n)s(n) + w(n); & H_1, \end{cases} \qquad (5.1)$$

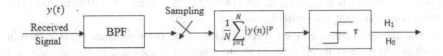

**Figure 5.1** Block diagram of generalized energy detector.

where $h(n), s(n)$ and $w(n)$ correspond to $n^{th}$ sample of the complex fading channel gain, PU signal and the noise, respectively with $n = 1, 2, \ldots, N$. The signal and noise samples are independent and identically distributed (iid) with $s(n) \sim \mathscr{CN}(0, \sigma_s^2)$ and $w(n) \sim \mathscr{CN}(0, \sigma_w^2)$. Complex Gaussian signal assumption is valid, for example, in an orthogonal frequency-division multiplexing signal having a large number of subcarriers [188, 211], in frequency-shift keying signals that can be reasonably approximated as Gaussian process due to the complex time structure.

Here, the notation $\mathscr{CN}(\bar{x}, \sigma_x^2)$ denotes complex Gaussian distribution with mean $\bar{x}$ and variance $\sigma_x^2$. The signal and noise are statistically independent of each other. The hypotheses $H_0$ and $H_1$ correspond to free and occupied primary channel, respectively.

The block diagram of the generalized energy detector is shown in Fig. 5.1. Like ED, the received signal $y(t)$ is passed through a bandpass filter to take out the band of interest, which is then sampled. In CED, the absolute values of received samples are squared and summed over the number of collected samples and then compared with the predetermined threshold to decide on the presence and the absence of the PU. In the generalized energy detector (GED) [47], the squaring operation is replaced by an arbitrary positive value of $p$. Hence, the received signal decision statistic for GED is given by

$$T = \frac{1}{N} \sum_{n=1}^{N} |y(n)|^p, \quad (5.2)$$

where, $N$ is the number of collected samples (sample size) and $p > 0$ is an arbitrary constant. As already discussed in Chapter 4, the performance of the GED can be measured in terms of $P_F$ and $P_D$.

$$P_F = Pr\{T > \tau | H_0\} \text{ and } P_D = Pr\{T > \tau | H_1\}, \quad (5.3)$$

where, $\tau$ is the decision threshold and $Pr\{\cdot\}$ is the probability operator.

## 5.2 NOISE UNCERTAINTY MODEL

The characterization of AWGN, i.e., $w(n)$, in Eq. (5.1) depends on its variance. In general, for many detection methods, it is assumed that the true noise variance at the input of SU is known a priori. These methods use this knowledge to choose a threshold in detecting the presence or the absence of the PU signal. However, in practice, the noise variance may vary over time and location, thus giving rise to a phenomenon called noise uncertainty [189, 232], which makes it difficult to obtain exact noise variance at a particular time and location.

The average value, i.e., the expected value of the noise variance $\hat{\sigma}_w^2$ is known in practice. As already mentioned, let the true noise variance at a particular time and location be $\sigma_w^2$ which may vary from the average noise variance giving rise to noise uncertainty. Using this, the noise uncertainty factor $\beta$ is defined as $\beta = \frac{\hat{\sigma}_w^2}{\sigma_w^2}$. Note that $\beta$ is a random variable since $\sigma_w^2$ is random. Let the upper bound on the noise uncertainty be $L$ dB, which is defined as

$$L = sup\{10 log_{10} \beta\}. \quad (5.4)$$

In this work, we assume that $\beta$ in dB is uniformly distributed in the range $[-L, L]$ [189], which implies $\beta$ is restricted in the range $[10^{\frac{-L}{10}}, 10^{\frac{L}{10}}]$. The pdf of $\beta$ can be obtained by using simple transformation of random variable as

$$f_\beta(x) = \begin{cases} 0, & x < 10^{\frac{-L}{10}} \\ \frac{5}{[ln(10)]Lx}, & 10^{\frac{-L}{10}} < x < 10^{\frac{L}{10}} \\ 0, & x > 10^{\frac{L}{10}} \end{cases} \quad (5.5)$$

where, $ln(z)$ is the natural logarithm of $z$.

## 5.3 SNR WALL FOR AWGN CHANNEL

For a given SNR $> 0$, if there exists a threshold for which

$$\lim_{N \to \infty} \bar{P}_F = 0 \text{ and } \lim_{N \to \infty} \bar{P}_D = 1, \quad (5.6)$$

then the sensing scheme is considered as unlimitedly reliable [232]. In other words, if the channel is sensed for sufficiently long time, i.e., $N \to \infty$, one can achieve desired target probability of false alarm and the probability of detection at any SNR level. However, this is possible only when there is no noise uncertainty. When there exists noise uncertainty, it is not possible to achieve this performance even with the use of unlimited sample size ($N$) below some SNR value [232]. The SNR value below which it is not possible to achieve unlimited reliability is defined as the SNR wall [232]. At least one of the conditions in Eq. (5.6) is not satisfied if the SNR falls below the SNR wall. However, when the SNR is above the SNR wall there exists a threshold for which both the conditions in Eq. (5.6) are satisfied.

In this section, we derive $\bar{P}_F$ and $\bar{P}_D$ in AWGN channel under noise uncertainty for no diversity, diversity and for CSS. Using them, we derive the expressions for SNR wall for each case. For relatively large $N$, using central limit theorem (CLT), the pdf of the decision statistic given in Eq. (5.2) can be modeled by Gaussian distribution [24, 97, 98, 206, 226] which can be represented by mean and variance only. Using these, the mean and variance of decision statistic are given as

$$\mu_0 = G_p \sigma_w^p, \quad \sigma_0^2 = \frac{K_p}{N} \sigma_w^{2p}, \text{ and}$$
$$\mu_1 = G_p (1+\gamma)^{\frac{p}{2}} \sigma_w^p, \quad \sigma_1^2 = \frac{K_p}{N}(1+\gamma)^p \sigma_w^{2p}, \quad (5.7)$$

where, $\mu_0$, $\sigma_0^2$ and $\mu_1$, $\sigma_1^2$ correspond to mean and variance of $T$ under $H_0$ and $H_1$, respectively. Here, $\gamma = |h|^2 \frac{\sigma_s^2}{\sigma_w^2}$ is the instantaneous SNR and since we are considering AWGN channel in this section, $h = 1$. Note that, in the remainder of this chapter, the term "SNR" without instantaneous means the average SNR. The $G_p$ and $K_p$ are given as

$$G_p = \Gamma\left(\frac{p+2}{2}\right), \text{ and } K_p = \Gamma(p+1) - \Gamma\left(\frac{p+2}{2}\right)^2. \quad (5.8)$$

Here, $\Gamma(a) = \int_0^\infty x^{a-1} e^{-x} dx$ represents the complete Gamma function [4, 6.1.1].

### 5.3.1 NO DIVERSITY

In this section, the probability of false alarm and detection are derived first without considering noise uncertainty and then extended to consider noise uncertainty. Using the means and variances given in Eq. (5.7), one can obtain $P_F$ and $P_D$ without considering noise uncertainty as

$$P_F = Q\left(\frac{\tau - \mu_0}{\sigma_0}\right) \text{ and}$$
$$P_D = Q\left(\frac{\tau - \mu_1}{\sigma_1}\right), \quad (5.9)$$

respectively, where, $Q(t) = \frac{1}{\sqrt{2\pi}} \int_t^\infty e^{-\left(\frac{x^2}{2}\right)} dx$.

When we consider noise uncertainty, the mean and the variance of the decision statistic are given by

$$\mu_{0,nu} = G_p \sigma_w^p, \quad \sigma_{0,nu}^2 = \frac{K_p}{N} \sigma_w^{2p}, \quad (5.10)$$

$$\mu_{1,nu} = G_p (1+\beta\tilde{\gamma})^{\frac{p}{2}} \sigma_w^p, \quad \sigma_{1,nu}^2 = \frac{K_p}{N}(1+\beta\tilde{\gamma})^p \sigma_w^{2p}, \quad (5.11)$$

where, $\mu_{0,nu}$, $\sigma_{0,nu}^2$ and $\mu_{1,nu}$, $\sigma_{1,nu}^2$ correspond to the mean and the variance under $H_0$ and $H_1$, respectively and $\tilde{\gamma} = \frac{\sigma_s^2}{\sigma_w^2}$ is the average SNR. The subscript $nu$ indicates that the mean and variances are under noise uncertainty.

Now, $P_F$ and $P_D$ for fixed uncertainty factor $\beta$ can be obtained by substituting means and variances from Eq. (5.10) and Eq. (5.11) in Eq. (5.9). The threshold $\tau$ is chosen as $\lambda \hat{\sigma}_w^p$ for GED, where $\lambda \geq 0$ is a constant. However, when there exists noise uncertainty, $\beta$ is a random variable. In this case, one can obtain the average $P_F$ and $P_D$, i.e., $\bar{P}_F$ and $\bar{P}_D$, by averaging the $P_F$ and the $P_D$ obtained for fixed $\beta$ over the pdf of $\beta$ given in Eq. (5.5). Since they are functions of random variable $\beta$, after carrying out mathematical simplification, $\bar{P}_F$ and $\bar{P}_D$ can be obtained as

$$\bar{P}_F = \int_a^b Q\left(\left(\lambda x^{\frac{p}{2}} - G_p\right)\sqrt{\frac{N}{K_p}}\right)\frac{5}{Lxln(10)}dx, \text{ and} \tag{5.12}$$

$$\bar{P}_D = \int_a^b Q\left(\frac{\lambda x^{\frac{p}{2}} - G_p(1+x\tilde{\gamma})^{\frac{p}{2}}}{(1+x\tilde{\gamma})^{\frac{p}{2}}}\sqrt{\frac{N}{K_p}}\right)\frac{5}{Lxln(10)}dx, \tag{5.13}$$

where, $a = 10^{\frac{-L}{10}}$ and $b = 10^{\frac{L}{10}}$. Since the $Q$ function itself is an integral, the closed form solutions of Eq. (5.12) and Eq. (5.13) are mathematically involved and hence the expressions are kept in the integral form only.

To derive the SNR wall, we need to take limits as $N \to \infty$ in the expressions given in Eq. (5.12) and Eq. (5.13). After applying the limit, the reduced expressions for $\bar{P}_F$ and $\bar{P}_D$ are given by Eq. (5.14) and Eq. (5.15), respectively.

$$\bar{P}_F = \begin{cases} 0, & \lambda \geq G_p(b)^{\frac{p}{2}} \\ \frac{5}{Lln(10)}\left[\ln\left(\max\left\{\min\left\{\left(\frac{G_p}{\lambda}\right)^{\frac{2}{p}}, b\right\}, a\right\}\right) - \ln(a)\right], & G_p(a)^{\frac{p}{2}} < \lambda < G_p(b)^{\frac{p}{2}} \\ 1, & \lambda \leq G_p(a)^{\frac{p}{2}} \end{cases}$$
(5.14)

$$\bar{P}_D = \begin{cases} 0, & \lambda \geq G_p(\tilde{\gamma}+b)^{\frac{p}{2}} \\ \frac{5}{Lln(10)}\left[\ln\left(\max\left\{\min\left\{\frac{1}{\left(\frac{G_p}{\lambda}\right)^{\frac{2}{p}}-\tilde{\gamma}}, b\right\}, a\right\}\right) - \ln(a)\right], & G_p(\tilde{\gamma}+a)^{\frac{p}{2}} < \lambda < G_p(\tilde{\gamma}+b)^{\frac{p}{2}} \\ 1. & \lambda \leq G_p(\tilde{\gamma}+a)^{\frac{p}{2}} \end{cases}$$
(5.15)

Here, we choose not to give the derivation for these expressions since the steps involved in deriving these expressions also appear in Section 5.3.2 for which the steps involved are given in Appendix C.1.

Since the threshold is chosen by setting the value of $\lambda$, we need to find $\lambda$ for which both the conditions given in Eq. (5.6) are satisfied. Using Eq. (5.14) and Eq. (5.15), $\lambda$ should be chosen as

$$G_p\left(10^{\frac{L}{10}}\right)^{\frac{p}{2}} \geq \lambda \geq G_p\left(\tilde{\gamma}+10^{\frac{-L}{10}}\right)^{\frac{p}{2}}, \tag{5.16}$$

which gives us the condition on $\tilde{\gamma}$ for unlimited reliability as

$$\tilde{\gamma} \geq 10^{\frac{L}{10}} - 10^{\frac{-L}{10}}. \qquad (5.17)$$

The equality sign in Eq. (5.17) gives the lowest SNR for which unlimited reliability can be achieved and hence gives the SNR wall. It can be seen from Eq. (5.17) that when there is no noise uncertainty, i.e., $L = 0$, it is possible to find the threshold for unlimited reliability for any $\tilde{\gamma} > 0$. Hence, under no noise uncertainty, the GED is unlimitedly reliable. However, when there exists noise uncertainty, i.e., $L > 0$, GED is not unlimitedly reliable. One can also see from Eq. (5.17) that the SNR wall is independent of the value of $p$.

### 5.3.2 pLC DIVERSITY

The block diagram of the pLC diversity scheme is shown in Fig. 5.2. In the pLC diversity scheme, the decision statistic $T$ obtained at all the diversity branches are added and scaled by the total number of diversity branches to obtain a new decision statistic. The final decision on primary occupancy is taken after comparing the decision statistic against a threshold $\tau$. The decision statistic using pLC diversity can be written as

$$T_{plc} = \frac{1}{P} \sum_{i=1}^{P} T_i, \qquad (5.18)$$

where, $P$ and $T_i$ represent the total number of diversity branches and the decision statistic obtained at the $i^{th}$ diversity branch, respectively.

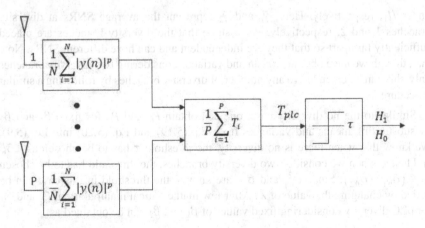

**Figure 5.2** Block diagram of pLC diversity technique.

To make the analysis easy to understand, we first consider a diversity scheme in which we have only two branches, i.e., $P = 2$, which is then extended to any number of diversity branches. Note that in practice, the noise uncertainties are different at each diversity branch, making us consider varying $\beta$'s at each branch. Let us also assume that the noise uncertainty associated with first branch is $\beta_1 = \frac{\hat{\sigma}_{w_1}^2}{\sigma_{w_1}^2}$ and that with branch two is $\beta_2 = \frac{\hat{\sigma}_{w_2}^2}{\sigma_{w_2}^2}$, where $\hat{\sigma}_{w_1}^2$ and $\hat{\sigma}_{w_2}^2$ are the average noise variances at diversity branches 1 and 2, respectively, which are known and $\sigma_{w_1}^2$ and $\sigma_{w_2}^2$ represent the true noise variances. With this setting, we have two noise uncertainties $\beta_1$ and $\beta_2$ which are uniformly distributed in the range $[-L_1, L_1]$ and $[-L_2, L_2]$, respectively. The noise uncertainty depends on calibration error, variations in thermal noise, and changes in low noise amplifier (LNA) gain, and hence different diversity branches can have different noise uncertainties.

Since the decision statistic obtained at two diversity branches, i.e., $T_1$ and $T_2$, are Gaussian distributed, the decision statistic $T_{plc}$ also follows Gaussian distribution with means and variances as

$$\mu_{0,nu} = \frac{G_p}{2}\left[\sigma_{w_1}^p + \sigma_{w_2}^p\right], \text{ and } \sigma_{0,nu}^2 = \frac{G_p K_p}{2^2 N}\left[\sigma_{w_1}^{2p} + \sigma_{w_2}^{2p}\right], \quad (5.19)$$

under $H_0$ and

$$\mu_{1,nu} = \frac{G_p}{2}\left[(1+\beta_1\tilde{\gamma}_1)^{\frac{p}{2}}\sigma_{w_1}^p + (1+\beta_2\tilde{\gamma}_2)^{\frac{p}{2}}\sigma_{w_2}^p\right],$$
$$\sigma_{1,nu}^2 = \frac{G_p K_p}{2^2 N}\left[(1+\beta_1\tilde{\gamma}_1)^p\sigma_{w_1}^{2p} + (1+\beta_2\tilde{\gamma}_2)^p\sigma_{w_2}^{2p}\right], \quad (5.20)$$

under $H_1$, respectively. Here, $\tilde{\gamma}_1$ and $\tilde{\gamma}_2$ represent the average SNRs at diversity branches 1 and 2, respectively. We assume that the diversity branches are placed sufficiently far apart so that they are independent and can have different SNRs. Note that, though we have obtained mean and variance considering two diversity branches only, this can be extended to any number of diversity branches by following a similar procedure.

Similar to the no diversity case, one can obtain $P_F$ and $P_D$ for fixed $\beta_1$ and $\beta_2$ by substituting means and variances from Eq. (5.19) and Eq. (5.20) into Eq. (5.9). We know that when there is no diversity, the threshold $\tau$ has to be chosen as $\lambda\hat{\sigma}_w^p$ and hence when we consider two diversity branches, the threshold has to be chosen as $\frac{\lambda}{2}\left(\hat{\sigma}_{w_1}^p + \hat{\sigma}_{w_2}^p\right)$. Since $\hat{\sigma}_{w_1}^2$ and $\hat{\sigma}_{w_2}^2$ are known, the threshold in this case can be varied by changing the value of $\lambda$. After few mathematical manipulation, $P_F$ and $P_D$ for pLC diversity considering fixed values of $\beta_1$ and $\beta_2$ can be obtained as

$$P_{F,plc} = Q\left(\frac{2\lambda\beta_1^{\frac{p}{2}}\beta_2^{\frac{p}{2}} - G_p\left(\beta_1^{\frac{p}{2}} + \beta_2^{\frac{p}{2}}\right)}{\sqrt{\beta_1^p + \beta_2^p}}\sqrt{\frac{N}{K_p}}\right) \text{ and } \quad (5.21)$$

$$P_{D,plc} = Q\left(\frac{2\lambda\beta_1^{\frac{p}{2}}\beta_2^{\frac{p}{2}} - \frac{G_p(1+\beta_1\tilde{\gamma}_1)^{\frac{p}{2}}}{\beta_2^{-\frac{p}{2}}} - \frac{G_p(1+\beta_2\tilde{\gamma}_2)^{\frac{p}{2}}}{\beta_1^{-\frac{p}{2}}}}{\sqrt{\frac{K_p}{N}}\sqrt{(1+\beta_1\tilde{\gamma}_1)^p\beta_2^p + (1+\beta_2\tilde{\gamma}_2)^p\beta_1^p}}\right), \quad (5.22)$$

respectively. Here, we have used $\beta_1 = \frac{\hat{\sigma}_{w_1}^2}{\sigma_{w_1}^2}$ and $\beta_2 = \frac{\hat{\sigma}_{w_2}^2}{\sigma_{w_2}^2}$.

We assume that the expected values of noise variance at both the diversity branches are one, i.e., $\hat{\sigma}_{w_1}^2 = \hat{\sigma}_{w_2}^2 = 1$. In the literature, when the noise uncertainty is not considered, the expected value of noise variance is considered the true noise variance. In our work, the expected value of variances is assumed to be 1 for mathematical simplicity. In [59, 60], the noise variance is assumed to be 1.

Now, the noise uncertainty being a random variable, one can obtain the average probability of false alarm ($\bar{P}_{F,plc}$) and detection ($\bar{P}_{D,plc}$) by averaging $P_{F,plc}$ in Eq. (5.21) and $P_{D,plc}$ in Eq. (5.22) over joint probability density function (jpdf) of random variables $\beta_1$ and $\beta_2$. Assuming that the two noise uncertainties $\beta_1$ and $\beta_2$ are independent, the jpdf of $\beta_1$ and $\beta_2$ can be given by

$$f_{\beta_1,\beta_2}(x,y) = \begin{cases} 0, & x < a_1, y < a_2, \\ \frac{25}{L_1 L_2 xy[\ln(10)]^2}, & a_1 < x < b_1, a_2 < y < b_2 \\ 0, & x > b_1, y > b_2, \end{cases} \quad (5.23)$$

where, $a_1 = 10^{\frac{-L_1}{10}}$, $b_1 = 10^{\frac{L_1}{10}}$, $a_2 = 10^{\frac{-L_2}{10}}$ and $b_2 = 10^{\frac{L_2}{10}}$. Here, $L_1$ and $L_2$ represent the upper bound on the noise uncertainty in dB at diversity branches 1 and 2, respectively, Using Eq. (5.23), $\bar{P}_{F,plc}$ is obtained as

$$\bar{P}_{F,plc} = \int_{a_1}^{b_1}\int_{a_2}^{b_2} Q\left(\frac{2\lambda(xy)^{\frac{p}{2}} - G_p\left(x^{\frac{p}{2}} + y^{\frac{p}{2}}\right)}{\sqrt{x^p + y^p}}\sqrt{\frac{N}{K_p}}\right)\frac{25}{[\ln(10)]^2 L_1 L_2 xy} dy dx. \quad (5.24)$$

Similarly, one can obtain $\bar{P}_{D,plc}$ as

$$\bar{P}_{D,plc} = \int_{a_1}^{b_1}\int_{a_2}^{b_2} Q\left(\frac{2\lambda(xy)^{\frac{p}{2}} - \frac{G_p(1+x\tilde{\gamma}_1)^{\frac{p}{2}}}{y^{-\frac{p}{2}}} - \frac{G_p(1+y\tilde{\gamma}_2)^{\frac{p}{2}}}{x^{-\frac{p}{2}}}}{\sqrt{\frac{K_p}{N}}\sqrt{(1+x\tilde{\gamma}_1)^p y^p + (1+y\tilde{\gamma}_2)^p x^p}}\right)\frac{25}{[\ln(10)]^2 L_1 L_2 xy} dy dx.$$

(5.25)

Now, to derive the SNR wall, we need to consider $N \to \infty$ in Eq. (5.24) and Eq. (5.25). After carrying out the mathematical simplifications, the $\bar{P}_{F,plc}$ can be reduced and the same is given by

$$\bar{P}_{F,plc} = \begin{cases} 0, & \lambda \geq \frac{G_p}{2}\left(e^{\frac{cp}{2}}+e^{\frac{dp}{2}}\right) \\ C_1\left[\frac{dp(A_1+A_2)}{2} - \frac{\ln(A-e^{-A_2})}{(A_2+\ln(A))^{-1}} + Li_2\left(\frac{A-e^{-A_2}}{A}\right) + \frac{\ln(A-e^{-A_1})}{(A_1+\ln(A))^{-1}} - Li_2\left(\frac{A-e^{-A_1}}{A}\right) + \frac{cdp^2}{2}\right], & \frac{G_p}{2}\left\{e^{\frac{-cp}{2}}+e^{\frac{-dp}{2}}\right\} < \lambda < \frac{G_p}{2}\left(e^{\frac{cp}{2}}+e^{\frac{dp}{2}}\right) \\ 1, & \lambda \leq \frac{G_p}{2}\left(e^{\frac{-cp}{2}}+e^{\frac{-dp}{2}}\right) \end{cases}$$

(5.26)

where, $Li_n(z)$ represents the polylogarithm [4]. Here also we avoid giving steps involved in the derivation since it involves steps similar to that given in APPENDIX C.1. In the expression given in Eq. (5.26), we have used $c = \ln(b_1)$, $d = \ln(b_2)$, $A = \frac{2\lambda}{G_p}$, $C_1 = \frac{100}{L_1 L_2 [p\ln(10)]^2}$, $R_1(z) = -\ln\left(A - e^{\frac{zp}{2}}\right)$, $A_1 = \max\left[\frac{-cp}{2}, R_1(-d)\right]$ and $A_2 = \min\left[\frac{cp}{2}, R_1(d)\right]$. Similarly, the expression for $\bar{P}_{D,plc}$ can be derived which can be approximated as given in Eq. (5.27),

$$\bar{P}_{D,plc} \approx \begin{cases} 0, & \lambda \geq \frac{G_p}{2}\left\{(\tilde{\gamma}_1+e^c)^{\frac{p}{2}}+(\tilde{\gamma}_2+e^d)^{\frac{p}{2}}\right\} \\ \frac{C_1 p^2}{4}(dU_2+I_1+2cd), & \frac{G_p}{2}\left\{(\tilde{\gamma}_1+e^{-c})^{\frac{p}{2}}+(\tilde{\gamma}_2+e^{-d})^{\frac{p}{2}}\right\} < \lambda < \frac{G_p}{2}\left\{(\tilde{\gamma}_1+e^c)^{\frac{p}{2}}+(\tilde{\gamma}_2+e^d)^{\frac{p}{2}}\right\} \\ 1, & \lambda \leq \frac{G_p}{2}\left\{(\tilde{\gamma}_1+e^{-c})^{\frac{p}{2}}+(\tilde{\gamma}_2+e^{-d})^{\frac{p}{2}}\right\} \end{cases}$$

(5.27)

where, we use $R_2(z) = -\ln\left(\left(A-(e^z+\tilde{\gamma}_2)^{\frac{p}{2}}\right)^{\frac{2}{p}} - \tilde{\gamma}_1\right)$, $U_2 = \max[-c, R_2(-d)]$, $U_3 = \min[c, R_2(d)]$,

$$aa = \left(A-(1+\tilde{\gamma})^{\frac{p}{2}}\right)^{\frac{2}{p}}, \quad \alpha = \frac{-\left[(1+\tilde{\gamma})^{\left(\frac{p}{2}-1\right)}\right]}{A-(1+\tilde{\gamma})^{\frac{p}{2}}},$$

$$I_1 = \frac{aa}{\alpha}\left(e^{-\alpha U_3}-e^{-\alpha U_2}\right)+(1+\tilde{\gamma}_2)(U_3-U_2)+dU_3.$$

The derivation for $\bar{P}_{D,plc}$ is given in Appendix C.1. Once again, we need to find $\lambda$ for which both the conditions in Eq. (5.6) are satisfied. Hence, we need to select $\lambda$ as

$$e^{\frac{cp}{2}}+e^{\frac{dp}{2}} \leq \lambda \geq (\tilde{\gamma}_1+e^{-c})^{\frac{p}{2}}+(\tilde{\gamma}_2+e^{-d})^{\frac{p}{2}}. \tag{5.28}$$

The lowest value of $\tilde{\gamma}_1$ and $\tilde{\gamma}_2$ for which this inequality holds can be found by equating the upper and the lower limits. We can find $\tilde{\gamma}_1$ in terms of $\tilde{\gamma}_2$ for which the upper limit is equal to the lower limit. With this, we get the condition on $\tilde{\gamma}_1$ for unlimited reliability as

$$\tilde{\gamma}_1 \geq \left[e^{\frac{cp}{2}}+e^{\frac{dp}{2}}-(\tilde{\gamma}_2+e^{-d})^{\frac{p}{2}}\right]^{\frac{2}{p}}-e^{-c}. \tag{5.29}$$

Note that, here, we have obtained $\tilde{\gamma}_1$ in terms of $\tilde{\gamma}_2$. However, one can obtain $\tilde{\gamma}_2$ in terms of $\tilde{\gamma}_1$. In Eq. (5.29), one needs to select $\tilde{\gamma}_2$ such that $\tilde{\gamma}_1 \geq 0$ giving us the condition on $\tilde{\gamma}_2$ as

$$0 \leq \tilde{\gamma}_2 \leq \left(e^{\frac{cp}{2}} + e^{\frac{dp}{2}} - e^{\frac{-cp}{2}}\right)^{\frac{2}{p}} - e^d. \tag{5.30}$$

The two conditions in Eq. (5.29) and Eq. (5.30) represent the conditions for SNR wall, when we consider the pLC diversity with two diversity branches. Looking at these expressions, one can draw the following conclusions.

- In the absence of noise uncertainty at the input of both the diversity branches, i.e., $L_1 = L_2 = 0$, from Eq. (5.29) and Eq. (5.30), we get $\tilde{\gamma}_1 > 0$ and $\tilde{\gamma}_2 > 0$. This indicates that one can find a threshold for which both the conditions in Eq. (5.6) are satisfied at any SNR greater than zero. This means that the sensing scheme is unlimitedly reliable when there is no noise uncertainty.
- Let us consider that both the branches have same noise uncertainties, i.e., $L_1 = L_2$, and SNR, i.e., $\tilde{\gamma}_1 = \tilde{\gamma}_2$. Substituting $L_1 = L_2 = L$ and $\tilde{\gamma}_1 = \tilde{\gamma}_2 = \tilde{\gamma}$ in Eq. (5.29) we get

$$(\tilde{\gamma} + e^{-c})^{\frac{p}{2}} \geq e^{\frac{cp}{2}} + e^{\frac{dp}{2}} - (\tilde{\gamma} + e^{-d})^{\frac{p}{2}}. \tag{5.31}$$

Since $L_1 = L_2 = L$, we get $c = d$. Substituting for $c = d$ and solving for $\tilde{\gamma}$, we get

$$\tilde{\gamma} \geq 10^{\frac{L}{10}} - 10^{\frac{-L}{10}}. \tag{5.32}$$

Hence, we get the same SNR wall expression as in Eq. (5.17) that is, the SNR wall is the same as in the case of no diversity case. This indicates that there is no improvement in terms of the SNR wall. We also observe that the SNR wall is independent of the value of $p$.
- In the scenario when both the diversity branches have same SNRs, i.e., $\tilde{\gamma}_1 = \tilde{\gamma}_2 = \tilde{\gamma}$, and different noise uncertainties, i.e., $L_1 \neq L_2$, the SNR walls remain almost the same for all $p$. For example, with $L_1 = 1\ dB$ and $L_2 = 0.5\ dB$ in Eq. (5.29), we get $\tilde{\gamma} = 0.3472$ when $p = 1$ and $\tilde{\gamma} = 0.3677$ as $p \to \infty$ which are approximately the same.
- The advantage of using diversity lies in the fact that SNRs determine the SNR wall at different diversity branches, i.e., $\tilde{\gamma}_1$ and $\tilde{\gamma}_2$. Hence, even if the SNR is low at one diversity branch, and the other has sufficiently high SNR, we could still satisfy conditions to achieve unlimitedly reliable sensing. For example, with $L_1 = L_2 = 1\ dB$ and $p = 2$, the unlimited reliability can be obtained if one of the diversity branches has SNR of 0.3, and the other has SNR of 0.6262.
- The SNR wall depends on the selected value of $p$, i.e., with an increasing value of $p$, the SNR wall decreases. This is demonstrated with the help of Fig. 5.3, where we plot $p$ VS. $\tilde{\gamma}_1$ for fixed value of $\tilde{\gamma}_2 = 0.1$. We can see that with an increasing value of $p$ the value of $\tilde{\gamma}_1$ goes down, showing improvement in terms of SNR wall.

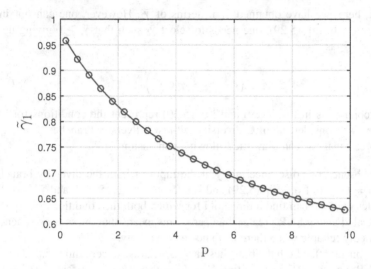

**Figure 5.3** $p$ VS. $\tilde{\gamma}_1$ for $\tilde{\gamma}_2 = 0.1$, $L_1 = L_2 = 1\ dB$.

- For a fixed value of $\tilde{\gamma}_2$ the smallest value of $\tilde{\gamma}_1$ that can be achieved as we increase $p$ can be found by setting $p \to \infty$ in Eq. (5.29). The value of $\tilde{\gamma}_1$ obtained will now depend on the values of $L_1$ and $L_2$. Note that when we consider $L_1 > L_2$, the term $e^{\frac{dp}{2}} - (\tilde{\gamma}_2 + e^{-d})^{\frac{p}{2}}$ becomes very small when compared to $e^{\frac{cp}{2}}$ as $p \to \infty$. Under this condition we get

$$\tilde{\gamma}_1 \geq e^c - e^{-c} = 10^{\frac{L_1}{10}} - 10^{\frac{-L_1}{10}}. \qquad (5.33)$$

Similarly, if $L_2 > L_1$, we get

$$\tilde{\gamma}_1 \geq e^d - e^{-c} = 10^{\frac{L_2}{10}} - 10^{\frac{-L_1}{10}}. \qquad (5.34)$$

Finally, if $L_1 = L_2 = L$, we get

$$\tilde{\gamma}_1 \geq = 10^{\frac{L}{10}} - 10^{\frac{-L}{10}}. \qquad (5.35)$$

For example, with $L_1 = 1\ dB$, $L_2 = 0.5\ dB$ and $\tilde{\gamma}_2 = 0.1$, we require $\tilde{\gamma}_1 \geq 0.4646$ as $p \to \infty$. This shows that with $\tilde{\gamma}_2 = 0.1$, one can have $\tilde{\gamma}_1$ as small as 0.4646 and still get the unlimited performance.

Now the derivation for $P_{F,plc}$ and $P_{D,plc}$ for the two diversity branches given in Eq. (5.21) and Eq. (5.22), respectively, can be extended to $P$ number of branches by selecting the threshold $\tau$ as $\tau = \frac{\lambda}{p}(\hat{\sigma}_{w_1}^p + \hat{\sigma}_{w_2}^p + \cdots + \hat{\sigma}_{wp}^p)$. Following a similar procedure, $P_{F,plc}$ and $P_{D,plc}$ for $P$ number of diversity branches and fixed values of noise uncertainty factors $\beta_1, \beta_2, \ldots, \beta_P$ are obtained as

$$P_{F,plc} = Q\left(\frac{P\lambda \prod_{i=1}^{P} \beta_i^{\frac{p}{2}} - G_p \sum_{i=1}^{P} \prod_{j=1, j\neq i}^{P} \beta_j^{\frac{p}{2}}}{\sqrt{\sum_{i=1}^{P} \prod_{j=1, j\neq i}^{P} \beta_j^{p}}} \sqrt{\frac{N}{K_p}}\right), \quad (5.36)$$

$$P_{D,plc} = Q\left(\frac{P\lambda \prod_{i=1}^{P} \beta_i^{\frac{p}{2}} - \sum_{i=1}^{P} G_p(1+\beta_i\tilde{\gamma}_i)^{\frac{p}{2}} \prod_{j=1, j\neq i}^{P} \beta_j^{\frac{p}{2}}}{\sqrt{\frac{K_p}{N}}\sqrt{\sum_{i=1}^{P}(1+\beta_i\tilde{\gamma}_i)^{p} \prod_{j=1, j\neq i}^{P} \beta_j^{p}}}\right), \quad (5.37)$$

respectively. Averaging the $P_{F,plc}$ and $P_{D,plc}$ in Eq. (5.36) and Eq. (5.37) over jpdf of $\beta_1, \beta_2, \ldots, \beta_P$, we get $\bar{P}_{F,plc}$ and $\bar{P}_{D,plc}$. To derive the SNR wall in this case, one can follow the procedure similar to two diversity branches. To keep it simple, we derive the SNR wall by considering $p = 2$. In this case, to achieve unlimited reliability, we need to select $\lambda$ as

$$10^{\frac{L_1}{10}} + 10^{\frac{L_2}{10}} + \cdots + 10^{\frac{L_P}{10}} \leq \lambda \leq \tilde{\gamma}_1 + 10^{\frac{-L_1}{10}} + \tilde{\gamma}_2 + 10^{\frac{-L_2}{10}} + \cdots + \tilde{\gamma}_P + 10^{\frac{-L_P}{10}}. \quad (5.38)$$

Using this, the SNR wall can be given as

$$\sum_{i=0}^{P} \tilde{\gamma}_i = \sum_{i=0}^{P} 10^{\frac{L_i}{10}} - \sum_{i=0}^{P} 10^{\frac{-L_i}{10}}, \quad (5.39)$$

where, $L_i$ and $\tilde{\gamma}_i$ represent the upper bounds on the noise uncertainty and the average SNR at the $i^{th}$ diversity branch with $i = 1, 2, \ldots, P$. One can see from Eq. (5.39) that, we need to set the combined SNR, i.e., $\tilde{\gamma}_1 + \tilde{\gamma}_2 + \cdots + \tilde{\gamma}_P$, to achieve unlimited performance. The conclusions that we draw for the case of $P = 2$ can be directly applied to this general case as well.

## 5.3.3 pLS DIVERSITY

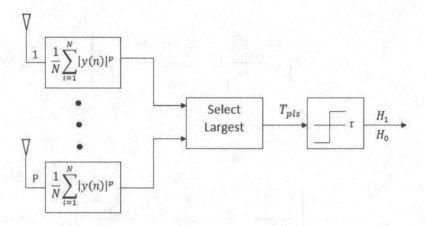

**Figure 5.4** Block diagram of pLS diversity technique.

The schematic representation of the pLS diversity technique is shown in Fig. 5.4. In pLS diversity scheme, the new decision statistic is obtained as the maximum of the decision statistics obtained at the diversity branches which is given as $T_{pls} = max\{T_1, T_2, \ldots, T_P\}$, where $T_1, T_2, \ldots, T_P$ represent the decision statistics obtained at $M$ diversity branches. The decision on the occupancy of the PU is then taken after comparing $T_{pls}$ against the threshold $(\tau)$.

Since we assume that the diversity branches receive independent decision statistics, the average probability of false alarm can be expressed as

$$\bar{P}_{F,pls} = 1 - \prod_{i=1}^{P}(1 - \bar{P}_{F_i}), \qquad (5.40)$$

where, $\bar{P}_{F_i}$ corresponds to the average probability of false alarm at the $i^{th}$ diversity branch which can be obtained by using Eq. (5.14). Similarly, the average probability of detection in this case can be obtained as

$$\bar{P}_{D,pls} = 1 - \prod_{i=1}^{P}(1 - \bar{P}_{D_i}), \qquad (5.41)$$

where, $\bar{P}_{D_i}$ corresponds to the average probability of detection at the $i^{th}$ diversity branch. The $\bar{P}_{D,pls}$ can be obtained by using Eq. (5.15) in Eq. (5.41).

In this case also, we use $P = 2$ to derive the SNR wall and then extend the analysis to general case. Substituting $P = 2$ in Eq. (5.40), $\bar{P}_{F,pls}$ can be obtained as

$$\bar{P}_{F,pls} = 1 - (1 - \bar{P}_{F_1})(1 - \bar{P}_{F_2}), \qquad (5.42)$$

where, $\bar{P}_{F_1}$ and $\bar{P}_{F_2}$ are the average probability of false alarm associated with branches 1 and 2, respectively. It is clear from Eq. (5.42) that to achieve $\lim_{N\to\infty}\bar{P}_{F,pls} =$

0, both $\bar{P}_{F_1}$ and $\bar{P}_{F_2}$ must be 0. Let $L_1$ and $L_2$ be upper bound on the uncertainties associated with branches 1 and 2, respectively.

To set $\bar{P}_{F_1} = 0$, using Eq. (5.14) it is clear that $\lambda$ should be selected as

$$\lambda \geq G_p \left(10^{\frac{L_1}{10}}\right)^{\frac{p}{2}}. \tag{5.43}$$

Similarly, to set $\bar{P}_{F_2} = 0$, $\lambda$ has to satisfy

$$\lambda \geq G_p \left(10^{\frac{L_2}{10}}\right)^{\frac{p}{2}}. \tag{5.44}$$

Since we need both $\bar{P}_{F_1}$ and $\bar{P}_{F_2}$ as 0, $\lambda$ has to satisfy

$$\lambda \geq max\left\{G_p \left(10^{\frac{L_1}{10}}\right)^{\frac{p}{2}}, G_p \left(10^{\frac{L_2}{10}}\right)^{\frac{p}{2}}\right\}. \tag{5.45}$$

After substituting $P = 2$ in Eq. (5.41), we get $\bar{P}_{D,pls}$ as

$$\bar{P}_{D,pls} = 1 - (1 - \bar{P}_{D_1})(1 - \bar{P}_{D_2}), \tag{5.46}$$

where, $\bar{P}_{D_1}$ and $\bar{P}_{D_2}$ are the average probability of detection associated with branch 1 and 2, respectively. We can see from Eq. (5.46) that to achieve $\lim_{N \to \infty} \bar{P}_{D,pls} = 1$, $\bar{P}_{D_1}$ or $\bar{P}_{D_2}$ must be 1. To set $\bar{P}_{D_1} = 1$, $\lambda$ has to be selected as

$$\lambda \leq G_p \left(10^{\frac{-L_1}{10}} + \tilde{\gamma}_1\right)^{\frac{p}{2}}. \tag{5.47}$$

Similarly, to set $\bar{P}_{D_2}$ to 1,

$$\lambda \leq G_p \left(10^{\frac{-L_2}{10}} + \tilde{\gamma}_2\right)^{\frac{p}{2}}. \tag{5.48}$$

Since any one of the conditions given in Eq. (5.47) and Eq. (5.48) must be satisfied to get $\bar{P}_{D,pls} = 1$, they can be written in compact form as

$$\lambda \leq max\left\{G_p \left(10^{\frac{-L_1}{10}} + \tilde{\gamma}_1\right)^{\frac{p}{2}}, G_p \left(10^{\frac{-L_2}{10}} + \tilde{\gamma}_2\right)^{\frac{p}{2}}\right\}. \tag{5.49}$$

In order to see the implications of the conditions given in Eq. (5.45) and Eq. (5.49), let us consider $L_1 > L_2$. In this case, the conditions on $\tilde{\gamma}_1$ and $\tilde{\gamma}_2$ can be given by

$$\tilde{\gamma}_1 \geq 10^{\frac{L_1}{10}} - 10^{\frac{-L_1}{10}} \text{ OR } \tilde{\gamma}_2 \geq 10^{\frac{L_1}{10}} - 10^{\frac{-L_2}{10}}. \tag{5.50}$$

From this, the SNR wall for pLS diversity is obtained by considering the equality sign in Eq. (5.50). Hence, to achieve unlimited reliability, SNR at the input of any one branch has to be $\geq$ its respective SNR wall. To understand this let us take $L_1 = 0.5\,dB$ and $L_2 = 0.3\,dB$. Substituting in Eq. (5.50), we get $\tilde{\gamma}_1 = 0.2308$ and $\tilde{\gamma}_2 = 0.1888$.

Therefore, to achieve unlimited reliability we must have either $\tilde{\gamma}_1 \geq 0.2308$ or $\tilde{\gamma}_2 \geq 0.1888$. One can also see from Eq. (5.50) that unlike pLC case, the SNR wall, in this case, is independent of $p$. Another advantage of pLS diversity is that any one branch should have SNR $\geq$ its SNR wall, and hence even if the other branch is experiencing worst channel conditions, one can achieve the unlimited reliability with sufficiently high SNR at another branch. Following a similar procedure, the analysis for $P = 2$ can be extended to any $P$ and is given as

$$\tilde{\gamma}_i \geq 10^{\frac{L^+}{10}} - 10^{\frac{-L_i}{10}}, \text{ for } i = 1, 2, \ldots, P, \tag{5.51}$$

where, $L^+ = \max[L_1, L_2, \ldots, L_P]$. In this case, to achieve unlimited reliability, any one among $P$ conditions in Eq. (5.51) must be satisfied.

### 5.3.4 CSS WITH HARD COMBINING

Until now, we discussed deriving the SNR walls when we consider diversity. We now discuss the same when CSS is used where multiple CSUs collaborate by sharing their information to detect the presence or the absence of the PU. Let us consider that there is $M$ number of independent CSUs, and each one of them receives $N$ samples during the observation interval. We denote the true and average noise variance at the $i^{\text{th}}$ CSU as $\sigma_{w_i}^2$ and $\hat{\sigma}_{w_i}^2$, respectively, where $i = 1, 2, \ldots, M$. Note that we are not considering diversity reception for CSUs. The decision statistic obtained at the $i^{\text{th}}$ CSU is denoted as $T_i$. Once again we assume that the noise uncertainty factor $\beta_i$ at the $i^{\text{th}}$ CSU is uniformly distributed in the interval $[-L_i, L_i]$. In the case of hard decision combining, all the CSUs take independent decisions regarding PU's occupancy and send the results as ON/OFF to the fusion center (FC). The FC then takes the final decision considering all the received decisions. Let $\bar{Q}_F$ and $\bar{Q}_D$ denote the average probability of false alarm and detection at the FC, respectively. When using CSS, one has to modify the SNR wall definition, which is given as follows. Given the different SNRs ($\tilde{\gamma}_i > 0$) at the SUs, $i = 1, 2, \ldots, M$, if there exists a threshold for which

$$\lim_{N \to \infty} \bar{Q}_F = 0 \text{ and } \lim_{N \to \infty} \bar{Q}_D = 1, \tag{5.52}$$

then the sensing scheme is considered as unlimitedly reliable [232]. We require $\bar{Q}_F$ and $\bar{Q}_D$ as $N \to \infty$ in order to derive the SNR wall and to derive the same, first we need to obtain average probability of false alarm ($\bar{P}_{F_i}$) and detection ($\bar{P}_{D_i}$) for the $i^{\text{th}}$ CSU considering $N \to \infty$. Using Eq. (5.12) and Eq. (5.13), $\bar{P}_{F_i}$ and $\bar{P}_{D_i}$ for $N \to \infty$ can be written as

$$\bar{P}_{F_i} = \begin{cases} 0, & \lambda \geq G_p(b_i)^{\frac{p}{2}} \\ \frac{5}{L_i \ln(10)} \left[ \ln \left( \max \left\{ \min \left\{ \left( \frac{G_p}{\lambda} \right)^{\frac{2}{p}}, b_i \right\}, a_i \right\} \right) - \ln(a_i) \right], & G_p(a_i)^{\frac{p}{2}} < \lambda < G_p(b_i)^{\frac{p}{2}} \\ 1, & \lambda \leq G_p(a_i)^{\frac{p}{2}} \end{cases} \tag{5.53}$$

# Generalized Energy Detector in the Presence of Noise Uncertainty and Fading

$$\bar{P}_{D_i} = \begin{cases} 0, & \lambda \geq G_p(\tilde{\gamma}_i + b_i)^{\frac{p}{2}} \\ \frac{5}{L_i \ln(10)} \left[ \ln\left( \max\left\{ \min\left\{ \frac{1}{\left(\frac{G_p}{\lambda}\right)^{\frac{2}{p}} - \tilde{\gamma}_i}, b_i \right\}, a_i \right\} \right) - \ln(a_i) \right], & G_p(\tilde{\gamma}_i + a_i)^{\frac{p}{2}} < \lambda < G_p(\tilde{\gamma}_i + b_i)^{\frac{p}{2}} \\ 1, & \lambda \leq G_p(\tilde{\gamma}_i + a_i)^{\frac{p}{2}} \end{cases}$$

(5.54)

where, $a_i = 10^{\frac{-L_i}{10}}$, $b_i = 10^{\frac{L_i}{10}}$ and $\tilde{\gamma}_i$ is the average SNR at the $i^{th}$ CSU. Using these, one can obtain $\bar{Q}_F$ and $\bar{Q}_D$, which can then be used to investigate the SNR wall for three combining rules, i.e., OR, AND, and $k$ out of $M$ combining rule.

### 5.3.4.1 OR Rule

In OR combining rule, the FC declares the PU as active whenever at least one of the CSUs reports the channel as occupied. Considering this, we first derive the SNR wall for $M = 2$ only and then extend the result to any number of CSUs. Let $L_1$ and $L_2$ be the upper bounds on the noise uncertainty factors and $\tilde{\gamma}_1$ and $\tilde{\gamma}_2$ be the SNRs at the two CSUs. In this case, $\bar{Q}_F$ and $\bar{Q}_D$ at the FC can be written as

$$\bar{Q}_F = \bar{P}_{F_1} + \bar{P}_{F_2} - \bar{P}_{F_1}\bar{P}_{F_2} \text{ and } \bar{Q}_D = \bar{P}_{D_1} + \bar{P}_{D_2} - \bar{P}_{D_1}\bar{P}_{D_2}. \quad (5.55)$$

From Eq. (5.55), it is clear that to satisfy $\lim_{N \to \infty} \bar{Q}_F = 0$, we need both $\bar{P}_{F_1}$ and $\bar{P}_{F_2}$ to be 0. Hence, using Eq. (5.53), one has to set the $\lambda$ at both the CSUs as

$$\lambda \geq G_p \left(10^{\frac{L_1}{10}}\right)^{\frac{p}{2}} \text{ AND } \lambda \geq G_p \left(10^{\frac{L_2}{10}}\right)^{\frac{p}{2}}. \quad (5.56)$$

The condition in Eq. (5.56) can be written in compact form as

$$\lambda \geq \max\left\{ G_p \left(10^{\frac{L_1}{10}}\right)^{\frac{p}{2}}, G_p \left(10^{\frac{L_2}{10}}\right)^{\frac{p}{2}} \right\}. \quad (5.57)$$

Similarly, to satisfy the condition $\lim_{N \to \infty} \bar{Q}_D = 1$, we see from Eq. (5.55) that $P_{D_1}$ or $P_{D_2}$ must be 1. Once again, using the Eq. (5.54), we need to set $\lambda$ as

$$\lambda \leq G_p \left(10^{\frac{-L_1}{10}} + \tilde{\gamma}_1\right)^{\frac{p}{2}} \text{ OR } \lambda \leq G_p \left(10^{\frac{-L_2}{10}} + \tilde{\gamma}_2\right)^{\frac{p}{2}}. \quad (5.58)$$

If we assume $L_1 > L_2$, then using the Eq. (5.57) and Eq. (5.58), $\lambda$ to be chosen for unlimited reliability should satisfy

$$G_p \left(10^{\frac{L_1}{10}}\right)^{\frac{p}{2}} \leq \lambda \leq G_p \left(10^{\frac{-L_1}{10}} + \tilde{\gamma}_1\right)^{\frac{p}{2}}, \text{ OR}$$
$$G_p \left(10^{\frac{L_1}{10}}\right)^{\frac{p}{2}} \leq \lambda \leq G_p \left(10^{\frac{-L_2}{10}} + \tilde{\gamma}_2\right)^{\frac{p}{2}}. \quad (5.59)$$

Using Eq. (5.59), the condition on $\tilde{\gamma}_1$ and $\tilde{\gamma}_2$ can be given as

$$\tilde{\gamma}_1 \geq 10^{\frac{L_1}{10}} - 10^{\frac{-L_1}{10}} \text{ OR } \tilde{\gamma}_2 \geq 10^{\frac{L_1}{2}} - 10^{\frac{-L_2}{10}}. \quad (5.60)$$

Therefore the SNR wall for the OR case is obtained by considering the equality condition in Eq. (5.60). To understand this, let us take $L_1 = 1 \ dB$ and $L_2 = 0.5 \ dB$. Substituting in Eq. (5.60), we get $\tilde{\gamma}_1 = 0.4646$ and $\tilde{\gamma}_2 = 0.3676$. Therefore one can achieve unlimited reliability if $\tilde{\gamma}_1 \geq 0.4646$ or $\tilde{\gamma}_2 \geq 0.3676$. One can also see from Eq. (5.60) that the SNR wall in this case is independent of $p$.

Following a similar procedure, the conditions given for the case of $M = 2$ in Eq. (5.60) can be extended to any $M$ as

$$\tilde{\gamma}_i \geq 10^{\frac{L^+}{10}} - 10^{\frac{-L_i}{10}}, \text{ for } i = 1, 2, \ldots, M, \quad (5.61)$$

where, $L^+ = \max\{L_1, L_2, \ldots, L_M\}$. In this case, to achieve unlimited reliability, any one among $M$ conditions in Eq. (5.61) must be satisfied.

### 5.3.4.2 AND Rule

Here, the FC declares the channel as occupied only when all the CSUs declare the PU channel as occupied. Similar to OR case, here also we first derive SNR wall by considering $M = 2$ and then extend it to any $M$. The $\bar{Q}_F$ and $\bar{Q}_D$ can be written as

$$\bar{Q}_F = \bar{P}_{F_1} \bar{P}_{F_2} \text{ and } \bar{Q}_D = \bar{P}_{D_1} \bar{P}_{D_2}. \quad (5.62)$$

It is clear from Eq. (5.62) that in order to satisfy the condition on $\bar{Q}_F$ in Eq. (5.52), either $\bar{P}_{F_1}$ or $\bar{P}_{F_2}$ must be 0. Hence, one has to select $\lambda$ as

$$\lambda \geq \min\left\{ G_p \left(10^{\frac{L_1}{10}}\right)^{\frac{p}{2}}, G_p \left(10^{\frac{L_2}{10}}\right)^{\frac{p}{2}} \right\}. \quad (5.63)$$

Similarly, to satisfy the condition on $\bar{Q}_D$, both $\bar{P}_{D_1}$ and $\bar{P}_{D_2}$ in Eq. (5.62) must be 1 and hence we need to set $\lambda$ as

$$\lambda \leq G_p \left(10^{\frac{-L_1}{10}} + \tilde{\gamma}_1\right)^{\frac{p}{2}} \text{ AND } \lambda \leq G_p \left(10^{\frac{-L_2}{10}} + \tilde{\gamma}_2\right)^{\frac{p}{2}}. \quad (5.64)$$

Once again assuming $L_1 > L_2$ and using Eq. (5.63) and Eq. (5.64), $\lambda$ has to be selected as

$$G_p \left(10^{\frac{L_2}{10}}\right)^{\frac{p}{2}} \leq \lambda \leq G_p \left(10^{\frac{-L_1}{10}} + \tilde{\gamma}_1\right)^{\frac{p}{2}} \text{ AND } G_p \left(10^{\frac{L_2}{10}}\right)^{\frac{p}{2}} \leq \lambda \leq G_p \left(10^{\frac{-L_2}{10}} + \tilde{\gamma}_2\right)^{\frac{p}{2}}. \quad (5.65)$$

Using this, $\tilde{\gamma}_1$ and $\tilde{\gamma}_2$ in this case should satisfy

$$\tilde{\gamma}_1 \geq 10^{\frac{L_2}{10}} - 10^{\frac{-L_1}{10}} \text{ AND } \tilde{\gamma}_2 \geq 10^{\frac{L_2}{10}} - 10^{\frac{-L_2}{10}}. \quad (5.66)$$

From this, the equality condition in the Eq. (5.66) gives us the SNR walls for the two CSUs. Once again considering $L_1 = 1 \ dB$ and $L_2 = 0.5 \ dB$, the unlimitedly reliable performance can be obtained if $\tilde{\gamma}_1 \geq 0.3277$ and $\tilde{\gamma}_2 \geq 0.2308$. Note that, in this case both the SNRs have to satisfy the inequality conditions. The conditions given in Eq. (5.66) can be extended to any number of $M$ and is given by Eq. (5.61) with $L^+ = \min\{L_1, L_2, \cdots, L_M\}$. Note that all the SNRs must be $\geq$ their respective SNR walls in order to achieve unlimited reliability.

### 5.3.4.3 $k$ Out of $M$ Combining Rule

In this rule, FC declares the channel as occupied when $k$ out of the total of $M$ CSUs report the PU channel as occupied. For this case, we first derive the SNR wall by considering $M = 3$ and $k = 2$, and then extend the result to the general case of $M$ and $k$. With this setting, $\bar{Q}_F$ and $\bar{Q}_D$ can be written as

$$\bar{Q}_F = \bar{P}_{F_1}\bar{P}_{F_2} + \bar{P}_{F_2}\bar{P}_{F_3} + \bar{P}_{F_1}\bar{P}_{F_3} - 2\bar{P}_{F_1}\bar{P}_{F_2}\bar{P}_{F_3}, \quad (5.67)$$

$$\bar{Q}_D = \bar{P}_{D_1}\bar{P}_{D_2} + \bar{P}_{D_2}\bar{P}_{D_3} + \bar{P}_{D_1}\bar{P}_{D_3} - 2\bar{P}_{D_1}\bar{P}_{D_2}\bar{P}_{D_3}. \quad (5.68)$$

Now for $\lim_{N \to \infty} \bar{Q}_F = 0$, we must have any of the two $\bar{P}_{F_i}$s, $i = 1,2,3$ must be 0 in Eq. (5.67). Therefore, $\lambda$ has to be selected such that

$$\lambda \geq \max\left\{G_p\left(10^{\frac{L_1}{10}}\right), G_p\left(10^{\frac{L_2}{10}}\right)\right\}, \text{ OR}$$

$$\lambda \geq \max\left\{G_p\left(10^{\frac{L_2}{10}}\right), G_p\left(10^{\frac{L_3}{10}}\right)\right\}; \text{ OR} \quad (5.69)$$

$$\lambda \geq \max\left\{G_p\left(10^{\frac{L_1}{10}}\right), G_p\left(10^{\frac{L_3}{10}}\right)\right\}.$$

To achieve the other condition of $\lim_{N \to \infty} \bar{Q}_D = 1$, using Eq. (5.68), any two $\bar{P}_{D_i}$s, for $i = 1,2,3$ must be 1 which is obtained by setting $\lambda$ as

$$\lambda \leq \min\left\{G_p\left(10^{\frac{-L_1}{10}} + \tilde{\gamma}_1\right)^{\frac{p}{2}}, G_p\left(10^{\frac{-L_2}{10}} + \tilde{\gamma}_2\right)^{\frac{p}{2}}\right\}, \text{ OR}$$

$$\lambda \leq \min\left\{G_p\left(10^{\frac{-L_1}{10}} + \tilde{\gamma}_1\right)^{\frac{p}{2}}, G_p\left(10^{\frac{-L_3}{10}} + \tilde{\gamma}_3\right)^{\frac{p}{2}}\right\}, \text{ OR} \quad (5.70)$$

$$\lambda \leq \min\left\{G_p\left(10^{\frac{-L_2}{10}} + \tilde{\gamma}_2\right)^{\frac{p}{2}}, G_p\left(10^{\frac{-L_3}{10}} + \tilde{\gamma}_3\right)^{\frac{p}{2}}\right\}.$$

In order to see the implications of these conditions, let us consider $L_1 > L_2 > L_3$. Using Eq. (5.69) and Eq. (5.70), the conditions on $\tilde{\gamma}_1$, $\tilde{\gamma}_2$ and $\tilde{\gamma}_3$ can be given by

$$\tilde{\gamma}_1 \geq 10^{\frac{L_2}{10}} - 10^{\frac{-L_1}{10}}, \tilde{\gamma}_2 \geq 10^{\frac{L_2}{10}} - 10^{\frac{-L_2}{10}}, \tilde{\gamma}_3 \geq 10^{\frac{L_2}{10}} - 10^{\frac{-L_3}{10}}. \quad (5.71)$$

Therefore, for $k = 2$ any two conditions given in Eq. (5.71) must be satisfied, in order to get unlimited reliability. Equality sign in Eq. (5.71) then gives us the SNR wall for 2 out of 3 rule. One can also see from Eq. (5.71) that the SNR wall in this case is independent of the value of $p$. As an example, let us take $L_1 = 1$ $dB$, $L_2 = 0.7$ $dB$ and $L_3 = 0.5$ $dB$. Substituting in Eq. (5.71), we get the SNR walls for 3 CSUs as $\tilde{\gamma}_1 = 0.3806$, $\tilde{\gamma}_2 = 0.3238$ and $\tilde{\gamma}_3 = 0.2836$. Therefore one can achieve unlimited reliability if any two of the SNRs at the CSUs are $\geq$ to their respective SNR wall values.

Following the similar procedure, the conditions given for the case of $M = 3$ in Eq. (5.71) can be extended to any $k$ out of $M$ CSUs and is given by Eq. (5.61) with

**Table 5.1**
**Comparison of SNR Walls for Hard Combining. Here,** $M = 3$, $L_1 = 1$ dB, $L_2 = 0.7$ dB, **and** $L_3 = 0.5$ dB.

| Decision Rule | $\tilde{\gamma}_1$ | $\tilde{\gamma}_2$ | $\tilde{\gamma}_3$ | k |
|---|---|---|---|---|
| OR | 0.4646 | 0.4077 | 0.3677 | 1 |
| AND | 0.3277 | 0.2708 | 0.2307 | 3 |
| 2 out of 3 | 0.3806 | 0.3238 | 0.2836 | 2 |

$L^+ = \min\{k \text{ largest from } (L_1, L_2, \ldots, L_M)\}$. For example, with $M = 3$, $k = 2$ and $L_1 > L_2 > L_3$ then $L^+ = L_2$ and we arrive at Eq. (5.71). Note that, to achieve unlimited reliability, any $k$ SNRs must be $\geq$ their respective SNR walls. In TABLE 5.1, we list the SNR wall under OR, AND, and $k$ out of $M$ combining rule when hard combining is used. We consider $k = 2$ for $k$ out of $M$ combining rule. Note that, the value of $k$ also represents the required number of SNRs to be $\geq$ their respective SNR walls at the CSUs to get the unlimited reliability. Looking at TABLE 5.1, one may notice that, though the SNR wall values that we get for OR combining rule are higher when compared to the other two rules, it requires only a one SNR to be $\geq$ the respective SNR wall value to achieve unlimited reliability. When AND combining rule is used, the SNR wall values are smallest, but we require all three SNRs $\geq$ their SNR wall values for achieving unlimited reliability. With $k$ out of $M$ combining rule, the SNR wall values lie between OR and AND combining rules, and any $k$ SNR values at the CSUs have to be $\geq$ their respective SNR wall values. The use of combining rules depends on the channel conditions of secondary users. If one of the SUs is experiencing good channel conditions so that the received SNR is high, then it is better to use OR combining rule. On the other hand, if all the SUs are experiencing low SNRs due to faded channel conditions, then one can employ AND combining rule. The K out N rule is useful when a few secondary users have good channel conditions.

### 5.3.5 CSS WITH SOFT COMBINING

We investigate the SNR wall for soft decision combining when equal gain combining (EGC) is used at the FC. Here, the decision on PU being ON/OFF is not taken by the CSUs. Instead, the decision statistics from all the CSUs are sent to the FC where they are added to obtain a new decision statistic and the decision is taken by FC based this. Let $T_i$ be the decision statistic at the $i^{th}$ CSU. Then, the new decision statistic at the FC is obtained as

$$T = \frac{1}{M}\sum_{i=1}^{M} T_i. \qquad (5.72)$$

We see that, the derivation for $\bar{Q}_F$ and $\bar{Q}_D$ in this case is same as that of $\bar{P}_{F,plc}$ and $\bar{P}_{D,plc}$ for pLC diversity given in Section 5.3.2, only the interpretation of the variables change. Specifically, $L_i$ and $\tilde{\gamma}_i$ used in Section 5.3.2 are now interpreted as the upper bound on the noise uncertainty and the average SNR at the $i^{th}$ CSU.

## 5.4 SNR WALL FOR FADING CHANNEL

In previous sections, we have derived SNR wall assuming that there are no fading effects in the channel and it is assumed to be AWGN. However, in real environment, the wireless channel undergoes fading effects. In this section, we first derive $\bar{P}_D$ under Nakagami fading and noise uncertainty for both no diversity and diversity cases. Since we know that $\bar{P}_F$ is independent of SNR, here we give the derivation for $\bar{P}_D$ only. We then discuss the SNR wall for each case.

### 5.4.1 NO DIVERSITY

The instantaneous SNR when we consider fading and no noise uncertainty is given by $\gamma = |h|^2 \frac{\sigma_s^2}{\sigma_w^2} = |h|^2 \frac{\sigma_s^2}{\hat{\sigma}_w^2}$ since $\sigma_w^2 = \hat{\sigma}_w^2$ in this case. This SNR at the input of the SU with Nakagami fading depends upon the random fading channel gain $\zeta$ as given in [27] as

$$f_\zeta(z) = \frac{m^m}{\Gamma(m)} z^{m-1} e^{-mz}, \; z \geq 0, \quad (5.73)$$

where, $m$ represents the Nakagami parameter. To obtain average probability of detection under Nakagami fading, i.e., $\bar{P}_D^{Nak}$, we need to replace $\tilde{\gamma}$ with $\tilde{\gamma}\zeta$ in Eq. (5.13) and average over the pdf of $\zeta$ given in Eq. (5.73) where $\bar{\gamma}$ is the average SNR when fading is considered. Note that $\tilde{\gamma}$ defined earlier in Section 5.3 is different from $\bar{\gamma}$. With this, $\bar{P}_D^{Nak}$ under Nakagami fading can be written as

$$\bar{P}_D^{Nak} = \int_0^\infty \int_a^b Q\left(\frac{\lambda x^{\frac{p}{2}} - G_p(1+\bar{\gamma}xz)^{\frac{p}{2}}}{(1+\bar{\gamma}xz)^{\frac{p}{2}}}\sqrt{\frac{N}{K_p}}\right) \frac{5}{Lx\ln(10)} \frac{m^m}{\Gamma(m)} z^{m-1} e^{-mz} dxdz. \quad (5.74)$$

Now, to obtain the SNR wall, we need to apply limit $N \to \infty$ to Eq. (5.74). In order to derive the expression, we define the following terms

$$R_3(z) = \left(\frac{1}{\bar{\gamma}}\left(\frac{\lambda}{G_p}\right)^{\frac{2}{p}} - \frac{1}{z\bar{\gamma}}\right), \quad (5.75)$$

$$U_5 = \max[0, R_3(a)] \text{ and } U_6 = \max[0, R_3(b)]. \quad (5.76)$$

After carrying out the mathematical simplifications, the $\bar{P}_D^{Nak}$ can be expressed as

$$\bar{P}_D^{Nak} = \frac{5}{L\ln(10)}\left[\left(1-\left(\frac{\lambda}{G_p}\right)^{\frac{2}{p}}\right)[P(m,mU_6)-P(m,mU_5)]-\ln(a)Q(m,mU_5)\right.$$
$$\left.+\bar{\gamma}[P(m+1,mU_6)-P(m+1,mU_5)]+\ln(b)Q(m,mU_6)\right], \quad (5.77)$$

where, $P(a,z)$ and $Q(a,z)$ represent the regularized lower and lower incomplete Gamma functions [4]. The derivation for the same is given in Appendix C.2.

Once the $\bar{P}_D^{Nak}$ is obtained, we need to obtain the condition on $\lambda$ such that both the conditions in Eq. (5.6) are satisfied. To get $\lim_{N\to\infty}\bar{P}_D^{Nak}=1$, using Eq. (5.74) we get the condition on $\lambda$ as

$$\lambda \le G_p\left(10^{\frac{-L}{10}}\right)^{\frac{p}{2}}. \quad (5.78)$$

Using Eq. (5.14), to set $\bar{P}_F = 0$, we need to select $\lambda$ as

$$\lambda \ge G_p\left(10^{\frac{L}{10}}\right)^{\frac{p}{2}}. \quad (5.79)$$

One can notice that, these two conditions cannot be satisfied simultaneously. For example, with $L = 1\ dB$, we need $\lambda \ge 1.2589$ to make $\bar{P}_F^{Nak} = 0$ and $\lambda \le 0.7943$ to achieve $\bar{P}_D^{Nak} = 1$. Such contradictory conditions arise because of the approximation of decision statistic as Gaussian. Hence, the expression given in Eq. (5.77) can never reach 1 for $\lambda \ge G_p\left(10^{\frac{L}{10}}\right)^{\frac{p}{2}}$. In such a situation, one can find the SNR wall numerically which can be done as follows. The condition given in Eq. (5.79) implies that to obtain $\lim_{N\to\infty}\bar{P}_F = 0$, we need to choose $\lambda \ge G_p\left(10^{\frac{L}{10}}\right)^{\frac{p}{2}}$. Now to maximize the probability of detection, we have to select the threshold as small as possible and hence $\lambda$ can be chosen as $G_p\left(10^{\frac{L}{10}}\right)^{\frac{p}{2}}$. With this, the SNR wall can be obtained by using this chosen threshold in Eq. (5.77) and finding the lowest value of SNR, i.e., $\bar{\gamma}$, for which $\bar{P}_D^{Nak}$ approximates 1. With $\lambda \ge G_p\left(10^{\frac{L}{10}}\right)^{\frac{p}{2}}$, the $\bar{P}_D^{Nak}$ never reaches 1 and hence we consider the SNR wall as that value of $\bar{\gamma}$ for which Eq. (5.77) approximates to 1, for example say 0.99. With $p = 2, L = 0.5\ dB$ and $m = 2$, we get the SNR wall as $\bar{\gamma} = 1.841$. For the same parameter settings, when we do not consider fading $\bar{\gamma} = 0.2308$.

### 5.4.2 pLC DIVERSITY

The $\bar{P}_{D,plc}^{Nak}$ considering two diversity branches can be obtained by averaging $\bar{P}_{D,plc}$ in Eq. (5.25) over the pdfs of $\zeta_1$ and $\zeta_2$ which is given by

$$\bar{P}_{D,plc}^{Nak} = C_2 \int_0^\infty \int_0^\infty \int_{a_1}^{b_1} \int_{a_2}^{b_2} Q\left( \frac{2\lambda(xy)^{\frac{p}{2}} - \frac{G_p(1+\bar{\gamma}_1 xz)^{\frac{p}{2}}}{y^{-\frac{p}{2}}} - \frac{G_p(1+\bar{\gamma}_2 yw)^{\frac{p}{2}}}{x^{-\frac{p}{2}}}}{\sqrt{\frac{K_p}{N}}\sqrt{(1+\bar{\gamma}_1 xz)^p y^p + (1+\bar{\gamma}_2 yw)^p x^p}} \right) \frac{1}{xy}(zw)^{m-1}e^{-m(z+w)}dydxdzdw.$$

(5.80)

where, $C_2 = \frac{25m^{2m}[\Gamma(m)]^{-2}}{L_1 L_2 \ln(10)^2}$. Here, $\bar{\gamma}_1$ and $\bar{\gamma}_2$ represent average SNRs under fading at diversity branches 1 and 2, respectively. Note that, we can reduce this expressions following the procedure given in Appendix C.2 for no diversity case but the final expression becomes too lengthy and complicated. Hence, in Appendix C.3, we give few initial steps of reduction to arrive at Eq. (5.81) which has three integrals to be reduced.

$$\bar{P}_{D,plc}^{Nak} = C_2 \int_0^\infty \int_0^\infty \int_{\ln(a_1)}^{\ln(b_1)} \left[ \max\left[ \min\left[ \ln(b_2), -\ln\left( \left( A - (\bar{\gamma}_1 z + e^{-t})^{\frac{p}{2}} \right)^{\frac{2}{p}} - \bar{\gamma}_2 w \right) \right], \right.$$

$$\left. \ln(a_2) \right] - \ln(a_2) \right] z^{m-1}e^{-mz}w^{m-1}e^{-mw} dt dz dw. \quad (5.81)$$

Here, we assume that both the diversity branches are experiencing independent fading. The analysis for $\bar{P}_{D,plc}^{Nak}$ can be extended to $P$ diversity branches by following a similar procedure, and in that case, the expression has $2P$ integrals where $P$ integrals are for averaging over noise uncertainty and another $P$ for averaging over the fading.

In this case also, the SNR wall has to be obtained numerically for the same reason explained in Section 5.4.1. Using Eq. (5.26), to achieve $\lim_{N\to\infty} \bar{P}_{F,plc} = 0$, we get the condition on $\lambda$ as

$$\lambda \geq \frac{G_p}{2}\left( \left(10^{\frac{L_1}{10}}\right)^{\frac{p}{2}} + \left(10^{\frac{L_2}{10}}\right)^{\frac{p}{2}} \right). \quad (5.82)$$

In this case, the SNR wall when both the diversity branches have the same SNR can be obtained by using $\bar{\gamma}_1 = \bar{\gamma}_2 = \bar{\gamma}$ in Eq. (5.81) and finding $\bar{\gamma}$ by considering $\lambda = (G_p/2)\left( \left(10^{\frac{L_1}{10}}\right)^{\frac{p}{2}} + \left(10^{\frac{L_2}{10}}\right)^{\frac{p}{2}} \right)$ for which $\bar{P}_{D,plc}^{Nak}$ attains 0.99. For example, with $L_1 = L_2 = 0.5\,dB$, $p = 2$ and $m = 2$, we get the SNR wall at the two diversity branches as $\bar{\gamma}_1 = \bar{\gamma}_2 = 0.67$. For no diversity case with $L = 0.5$, we get SNR wall as $\bar{\gamma} = 1.82$. We can observe that with the use of pLC diversity, we get improvement in terms of SNR wall. Finding SNR wall when we consider $\bar{\gamma}_1 \neq \bar{\gamma}_2$ is not possible unless we have additional constraints, since substituting $\lambda$ in Eq. (5.81) results in one equation with two unknowns which has infinite solutions. In this case, one can find the SNR wall by fixing the value of SNR at one of the diversity branches and then finding the other value numerically. For example, with the same parameter settings as given above and with $\bar{\gamma}_1 \neq \bar{\gamma}_2$, one combination of SNR for which unlimitedly reliable performance can be obtained is $\bar{\gamma}_1 = 0.5$ and $\bar{\gamma}_2 = 0.84$. One can follow the similar procedure to obtain the SNR wall when we consider more number of diversity branches.

### 5.4.3 pLS DIVERSITY

The average probability of detection under Nakagami fading, i.e., $\bar{P}_{D,pls}^{Nak}$, can be obtained by using Eq. (5.41) and Eq. (5.77). The average probability of false alarm remains the same as in the AWGN case given in Eq. (5.40).

In this case, one can select $\lambda$ as given in Eq. (5.45) and substitute the same in Eq. (5.77) to obtain the SNR walls for two diversity branches numerically. For example, with $L_1 = 0.5\ dB$, $L_2 = 0.3\ dB$, $m = 2$, and $p = 2$, we get the SNR walls at the two branches as $\bar{\gamma}_1 = 1.82$ and $\bar{\gamma}_2 = 1.71$. Once again, we see that any one of the two branches should have SNR above their respective SNR wall values to achieve unlimited reliability. A similar procedure can be followed to obtain the SNR wall for the general case of $P$ diversity branches. In this case, also, the SNR walls are independent of the value of $p$.

### 5.4.4 CSS WITH HARD COMBINING

We now consider the cooperative spectrum sensing scenario. In this case, to derive $\bar{Q}_F$ and $\bar{Q}_D$, we need to obtain the average probability of false alarm ($\bar{P}_{F_i}^{Nak}$) and detection ($\bar{P}_{D_i}^{Nak}$) at the $i^{th}$ CSU under Nakagami fading. Since $\bar{P}_{F_i}^{Nak}$ is independent of SNR, it remains the same as that of AWGN case given in Eq. (5.53). The derivation of $\bar{P}_{D_i}^{Nak}$ is the same as that of Eq. (5.77) which can be obtained by replacing $L$ and $\bar{\gamma}$ with $L_i$ and $\bar{\gamma}_i$, respectively, in Eq. (5.77) and is given as

$$\bar{P}_{D_i}^{Nak} = \frac{5}{L_i \ln(10)} \left[ \left(1 - \left(\frac{\lambda}{G_p}\right)^{\frac{2}{p}}\right) [P(m, mU_6) - P(m, mU_5)] - \ln(a_i) Q(m, mU_5) \right.$$
$$\left. + \bar{\gamma}_i [P(m+1, mU_6) - P(m+1, mU_5)] + \ln(b_i) Q(m, mU_6) \right], \quad (5.83)$$

where, $R_3(z)$, $U_5$ and $U_6$ are defined as follows

$$R_3(z) = \left(\frac{1}{\bar{\gamma}_i} \left(\frac{\lambda}{G_p}\right)^{\frac{2}{p}} - \frac{1}{z\bar{\gamma}_i}\right), \quad (5.84)$$

$$U_5 = \max[0, R_3(a_i)] \text{ and } U_6 = \max[0, R_3(b_i)]. \quad (5.85)$$

Here, we derive the SNR walls considering $k$ out of $M$ rule and the other two are discussed as the special cases. We first derive them by considering $M = 3$ and $k = 2$ and then extend it to general case. To find SNR wall, we first need to find $\lambda$ for which the conditions in Eq. (5.52) are satisfied which requires any of the two $\bar{P}_{F_i}$s be zero and $\bar{P}_{D_i}$s be one in Eq. (5.67) and Eq. (5.68), respectively. Using Eq. (5.53) and Eq. (5.83), to achieve $\bar{P}_{F_i} = 0$ and $\bar{P}_{D_i} = 1$, we need to select $\lambda$ as

$$\lambda \geq G_p \, (b_i)^{\frac{p}{2}} \text{ and } \lambda \leq G_p \, (a_i)^{\frac{p}{2}} \text{ for } i = 1, 2, 3, \tag{5.86}$$

respectively. Suppose we consider $L_1 > L_2 > L_3$. In that case, using Eq. (5.86), we need to set $\lambda \geq G_p(b_2)^{\frac{p}{2}}$ and $\lambda \leq G_p(a_2)^{\frac{p}{2}}$, respectively. However, these two conditions are contradictory and cannot be satisfied simultaneously. For example, with $p = 2$, $L_1 = 1 \, dB$, $L_2 = 0.7 \, dB$ and $L_3 = 0.5 \, dB$, we need to select $\lambda \geq 1.1749$ and $\lambda \leq 0.8511$. In such a situation, the expression given in Eq. (5.83) can never reach 1 for $\lambda \geq G_p(b_2)^{\frac{p}{2}}$ and we need to obtain the SNR walls numerically which can be done as follows. Note that, we always try to set $\lambda$ as small as possible since it maximizes the probability of detection. Hence, choosing $\lambda = G_p(b_2)^{\frac{p}{2}}$, the SNR wall for $i^{\text{th}}$ CSU can be found by using this threshold in Eq. (5.83) and by finding the lowest values of SNR, i.e., $\bar{\gamma}_i \ i = 1, 2, 3$, for which $\bar{P}_{D_i}^{Nak} \approx 1$. To do this, one may consider the SNR for which $\bar{P}_{D_i}^{Nak}$ reaches 0.99 (close to 1) as the SNR wall. For example, for the chosen uncertainties, we get the SNR walls as $\bar{\gamma}_1 = 2.88$, $\bar{\gamma}_2 = 2.62$ and $\bar{\gamma}_3 = 2.49$. Since $k = 2$, the SNRs at any two CSUs must be $\geq$ their respective SNR walls in order to achieve unlimited reliability. The case of $M = 3$ can be extended to any $M$ by choosing the threshold as $\lambda = G_p \left(10^{\frac{L^+}{10}}\right)^{\frac{p}{2}}$, where $L^+ = \min \{k \text{ largest from } (L_1, L_2, \ldots, L_M)\}$, and finding $\bar{\gamma}_i$ for which $\bar{P}_{D_i}^{Nak}$ approximates 1. In this case, any $k$ SNRs must be $\geq$ their respective SNR walls in order to achieve unlimited reliability. We next discuss OR and AND combining as the special cases of $k$ out of $M$ combining rule where the procedure to obtain SNR wall remains the same but the value of $L^+$ changes.

### 5.4.4.1 OR Combining

In OR combining, FC declares PU channel as occupied when at least one CSU declares channel as occupied. The SNR wall in this can be obtained by choosing $k = 1$ in $k$ out of $M$ rule. Hence, we get $L^+ = \max \{L_1, L_2, \ldots, L_M\}$. Similar to $k$ out of $M$ rule, by choosing the threshold as $\lambda = G_p \left(10^{\frac{L^+}{10}}\right)^{\frac{p}{2}}$ in Eq. (5.83) and finding the smallest $\bar{\gamma}_i$ for which we get $\bar{P}_{D_i}^{Nak} \approx 1$ gives us the SNR walls. Here, SNR at any one CSU needs to be $\geq$ its respective SNR wall to get unlimited reliability.

### 5.4.4.2 AND Combining

When AND combining is used, PU is considered as occupied when all the CSUs report it as occupied, and using $k = M$, one can get the SNR walls. with $k = M$ we get $L^+ = \min \{L_1, L_2, \ldots, L_M\}$. Following the procedure similar to the OR case, one can easily obtain the SNR walls numerically. Note that, since $k = M$, all CSUs must have SNRs $\geq$ their SNR walls in this case.

**Table 5.2**

Comparison of SNR Walls for Hard Combining under Nakagami Fading. Here, $M = 3$, $L_1 = 1\ dB$, $L_2 = 0.7\ dB$, and $L_3 = 0.5\ dB$.

| Decision Rule | $\bar{\gamma}_1$ | $\bar{\gamma}_2$ | $\bar{\gamma}_3$ | k |
|---|---|---|---|---|
| 2 out of 3 | 2.88 | 2.62 | 2.49 | 2 |
| OR | 3.85 | 3.67 | 3.58 | 1 |
| AND | 2.32 | 2.02 | 1.84 | 3 |

In TABLE 5.2, we list the SNR walls for different combining rules where the value of $k$ represents the required number of SNRs that are to be $\geq$ their respective SNR walls at the CSUs in order to get the unlimited reliability. One can see that the values of SNR walls obtained for the AND case are the lowest, but we need to have SNRs of all CSU above these values to achieve unlimited reliability. Although, the SNR walls for OR combining are high, only one CSU needs to have SNR $\geq$ its SNR wall. For $k$ out of $M$ rule, the SNR wall values lie between OR and AND combining rule. Another important point worth mentioning is that the SNR walls for hard combining are independent of $p$, i.e., we get the same SNR wall values as given in TABLE 5.2 for any value of $p$.

### 5.4.5 CSS WITH SOFT COMBINING

We need to obtain the SNR in this case numerically for the same reasons explained in Section 5.4.4. Once again, the process to obtain the SNR wall remains same as that for pLC diversity under Nakagami fading given in Section 5.4.2. The interpretation of $L_i$ and $\bar{\gamma}_i$ changes which are now interpreted as upper bound on the noise uncertainty and the average SNR under Nakagami fading at the $i^{th}$ CSU, respectively. The conclusions drawn in Section 5.4.2 are also valid in this case with $L_i$ and $\bar{\gamma}_i$ representing upper bound on the noise uncertainty and the average SNR under Nakagami fading at the $i^{th}$ CSU, respectively.

## 5.5 RESULTS AND DISCUSSION

In this section, we validate the theoretical analysis that is carried out using Monte Carlo simulations. For simulation, we generate both the PU signal and the noise as complex Gaussian, and the results are averaged over $10^5$ iterations. We first validate the expressions for SNR walls derived in Section 5.3 for the case of the AWGN channel. Then the validation of the same for fading case given in Section 5.4 is discussed.

### 5.5.1 SNR WALL FOR AWGN CASE

In this section, we consider experiments to verify the validity of SNR wall expressions derived in Sections 5.3 and 5.4. Since our significant contribution lies in deriving the SNR wall by considering diversity and CSS, we show the plots for validating the expressions for diversity and CSS cases only. We have already shown that the SNR wall is independent of $p$ for pLS diversity and CSS with hard combining. Therefore, we illustrate the SNR wall's validity by considering any $p$ for pLC diversity and by considering $p = 2$ for pLS diversity and CSS (hard combining). Note that one can easily obtain the plots to validate SNR walls for no diversity, pLS diversity, and CSS with hard combining with any $p$, but to avoid repetition of similar plots, we omitted those plots.

In Section 5.3.2 (Fig. 5.3), we conclude that with increase in the value of $p$, the SNR wall decreases. It is shown in Fig. 5.3 that by increasing the value of $p$, the SNR wall value decreases. We first validate this conclusion using Monte Carlo simulation. Here, we plot threshold $\tau$ VS. detection probabilities $\bar{P}_{F,plc}$, $\bar{P}_{D,plc}$, by selecting a large $N$ as required in establishing the SNR wall condition and by choosing $L_1 = L_2 = 1\ dB$ and $\tilde{\gamma}_2 = 0.1$. The infinite sample size is approximated by a very large $N$, i.e., $N = 10^7$. In Fig. 5.3, for $\tilde{\gamma}_2 = 0.1$, $L_1 = L_2 = 1\ dB$, we obtain the value of $\tilde{\gamma}_1$ for $p = 1$ and $p = 5$ as 0.8914 and 0.7153, respectively. Fig. 5.5a and Fig. 5.5b are plotted for $p = 1$ and $p = 5$, respectively. The vertical line in Fig. 5.5 gives us the threshold $\tau$ for which both the conditions in Eq. (5.6) are satisfied. This means that if we set $\tau \approx 0.9944$ for $p = 1$ and choose a very large value of $N$, we can achieve $\bar{P}_{F,plc} = 0$ and $\bar{P}_{D,plc} = 1$. In both these figures we can see that unlimited performance is indeed achieved but the required value of $\tilde{\gamma}_1$ (see Fig. 5.3) to achieve unlimited reliability for $p = 1$ is 0.8914 whereas for $p = 5$ it is 0.7153. We see that the value of SNR wall required at the branch 1 for $p = 5$ is 0.9958 $dB$ lower than that for $p = 1$. One can also obtain plots by choosing the values of $\tilde{\gamma}_1$ and $\tilde{\gamma}_2$ not satisfying the SNR wall conditions. In that case, it is not possible to find $\tau$ for which the conditions in Eq. (5.6) are satisfied. The plots in Fig. 5.5 are shown for the scenario where both the diversity branches have different SNRs. One can also validate the other conclusions drawn in Section 5.3.2.

In Fig. 5.6, we show the plots of $\tau$ VS. detection probabilities for pLS diversity with two branches. The simulation is carried out using $L_1 = 0.5\ dB$ and $L_2 = 0.3\ dB$. On substituting $L_1$ and $L_2$ in Eq. (5.50), we obtain the SNR walls as $\tilde{\gamma}_1 = 0.2308$ and $\tilde{\gamma}_2 = 0.1888$. As discussed in Section 5.3.3, any one of the SNRs at the two branches must be $\geq$ their respective SNR wall values to achieve unlimited reliability. Hence, we choose $\tilde{\gamma}_1 = 0.2$, which is below the SNR wall at the first diversity branch and $\tilde{\gamma}_2 = 0.1888$, which is equal to the SNR wall value at the second branch. It can be seen from Fig. 5.6 that by setting $\tau \approx 1.12$ and using high $N$, one can achieve unlimitedly reliable performance. The vertical line shows this in Fig. 5.6.In Fig. 5.6, we show the plots of $\tau$ VS. detection probabilities for pLS diversity with two branches. The simulation is carried out using $L_1 = 0.5\ dB$ and $L_2 = 0.3\ dB$. On substituting $L_1$ and $L_2$ in Eq. (5.50), we obtain the SNR walls as $\tilde{\gamma}_1 = 0.2308$ and $\tilde{\gamma}_2 = 0.1888$. As discussed in Section 5.3.3, any one of the SNRs at the two branches must be $\geq$

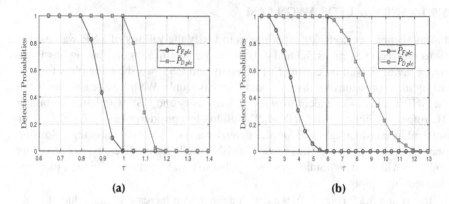

**Figure 5.5** Threshold ($\tau$) VS. detection probabilities for pLC diversity with two branches under AWGN channel using $L_1 = 1\ dB$, $L_2 = 1\ dB$, $\tilde{\gamma}_2 = 0.1$, $N = 10^7$ (a) for $p = 1$ and $\tilde{\gamma}_1 = 0.8914$ (b) for $p = 5$ and $\tilde{\gamma}_1 = 0.7153$.

their respective SNR wall values to achieve unlimited reliability. Hence, we choose $\tilde{\gamma}_1 = 0.2$, which is below the SNR wall at the first diversity branch and $\tilde{\gamma}_2 = 0.1888$, which is equal to the SNR wall value at the second branch. It can be seen from Fig. 5.6 that by setting $\tau \approx 1.12$ and using high $N$, one can achieve unlimitedly reliable performance. The vertical line shows this in Fig. 5.6.

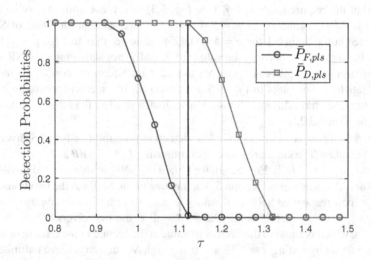

**Figure 5.6** Threshold ($\tau$) VS. detection probabilities for pLS diversity with two branches under AWGN channel using $L_1 = 0.5\ dB$, $L_2 = 0.3\ dB$, $\tilde{\gamma}_1 = 0.2$, $\tilde{\gamma}_2 = 0.1888$, $p = 2$, $N = 10^7$.

We next show the plots for CSS with one of the hard decision combining, i.e., $k$ out of $M$ combining rule. In Fig. 5.7, we demonstrate the SNR wall for $k$ out of $M$ combining rule and consider three CSUs, i.e., $M = 3$, having $L_1 = 1\ dB$, $L_2 = 0.7\ dB$ and $L_3 = 0.5\ dB$ with $k = 2$. Using these parameters in Eq. (5.71), we compute the SNR walls as $\tilde{\gamma}_1 = 0.3806$, $\tilde{\gamma}_2 = 0.3238$ and $\tilde{\gamma}_3 = 0.2836$. In Fig. 5.7, we show the plots by choosing $\tilde{\gamma}_1 = 0.2$ which is below the required value of SNR wall and $\tilde{\gamma}_2 = 0.3238$ and $\tilde{\gamma}_3 = 0.2836$ which are equal to their SNR walls. Since we have $k = 2$, and 2 out of 3 CSUs have the inputs with SNR $\geq$ their SNR walls, an unlimited operation is obtained. We can see from Fig. 5.7 that choosing a value of $\tau = 1.16$ (threshold corresponding to the vertical line) gives the unlimited reliability, i.e., choosing this threshold value with two of the three SNRs $\geq$ their SNR walls gives us $\bar{Q}_F = 0$ and $\bar{Q}_D = 1$. In order to reduce the repetition of the similar plots, here we give plots for $k$ out of $M$ combining rule only. One can show similar plots for OR as well as AND combining rules.

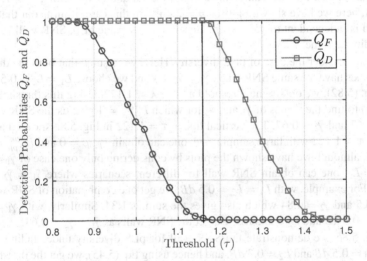

**Figure 5.7** Plots of $\tau$ VS. $\bar{Q}_F$ and $\bar{Q}_D$ for $k$ out of $M$ combining rule. Here, $N = 10^6$, $M = 3$, $L_1 = 1\ dB$, $L_2 = 0.7\ dB$, $L_3 = 0.5\ dB$, $p = 2$, $\tilde{\gamma}_1 = 0.2$, $\tilde{\gamma}_2 = 0.3238$, and $\tilde{\gamma}_3 = 0.2836$.

The SNR wall for CSS with soft combining is similar to the pLC diversity. Hence, one can consider plots given in Fig. 5.5 to understand the SNR walls for soft combining. For CSS with soft combining $\tilde{\gamma}_1$ and $\tilde{\gamma}_2$ represent the average SNRs at CSU 1 and 2, respectively, and $L_1$ and $L_2$ represent the noise uncertainty levels at CSU 1 and 2, respectively. We can see that for fixed value of SNR at CSU 2, i.e., $\tilde{\gamma}_2 = 0.1$, the required value of SNR at CSU 1, i.e., $\tilde{\gamma}_1$, in order to achieve unlimited reliability is 0.8914 for $p = 1$ and 0.7153 for $p = 5$ showing the improvement in SNR wall with increasing value of $p$.

## 5.5.2 SNR WALL FOR FADING CASE

In this section, we consider fading in addition to noise uncertainty and use Monte Carlo simulations to validate the SNR wall. We first consider the case of no diversity and then extend it to diversity and CSS. As already discussed, the SNR walls in this case have been obtained numerically, and one can find a threshold theoretically for which the probability of false alarm is 0 with an infinite sample size. We then need to find the SNR for which the probability of detection becomes 1 with this threshold to get the SNR wall.

In Fig. 5.8a, we show the plots for no diversity case. As discussed in Section 5.4.1, to achieve $\bar{P}_F^{Nak} = 0$, $\lambda$ should satisfy the condition given in Eq. (5.79). For $L = 0.5\,dB$ and $p = 2$, using Eq. (5.79), we get $\lambda \geq 1.122$. As already discussed, we always try to set the threshold as small as possible and hence we choose $\lambda = 1.122$. Since we assume $\hat{\sigma}_w^2 = 1$, $\lambda$ gives us the threshold. Using this $\lambda$ in Eq. (5.77) and finding $\bar{\gamma}$ for which $\bar{P}_D^{Nak} = 1$ gives us the SNR wall as $\bar{\gamma} = 1.82$. In Fig. 5.8a, the vertical line shows that we can achieve unlimited reliability by setting $\tau = 1.122$. Note that, here we have shown plot for $p = 2$ only. In this case, it turns out that the SNR wall is independent of value of $p$. Hence, for any value of $p$, SNR wall values will remain the same.

We next consider the case of pLC diversity. Here, we assume that both the diversity branches have the same SNRs, i.e., $\bar{\gamma}_1 = \bar{\gamma}_2 = \bar{\gamma}$ and we choose $L_1 = L_2 = 0.5\,dB$. Using Eq. (5.82), we obtain the threshold as $\tau = \lambda = 1.122$. Using this threshold in Eq. (5.81) to find the values of $\bar{\gamma}_1$ and $\bar{\gamma}_2$ for which $\bar{P}_{D,plc}^{Nak} = 1$ give us the SNR walls as $\bar{\gamma}_1 = 0.67$ and $\bar{\gamma}_2 = 0.67$. The vertical line at $\tau = 1.122$ in Fig. 5.8b shows that by choosing $\tau = 1.122$ with large sample size, one can obtain $\bar{P}_{F,plc} = 0$ and $\bar{P}_{D,plc}^{Nak} = 1$. Note that, although we have shown the plots by considering only one case of $\bar{\gamma}_1 = \bar{\gamma}_2$ and $L_1 = L_2$, one can obtain SNR wall for different scenarios where $\bar{\gamma}_1 \neq \bar{\gamma}_2$ and $L_1 \neq L_2$. For example, with $L_1 = L_2 = 0.5\,dB$, we get one combination of SNR walls as $\bar{\gamma}_1 = 0.5$ and $\bar{\gamma}_2 = 0.84$ which also gives the sum as 1.34. Similarly, with $\bar{\gamma}_1 = \bar{\gamma}_2$ with $L_1 = 0.5\,dB$ and $L_2 = 0.3\,dB$, we get the SNR walls as $\bar{\gamma}_1 = \bar{\gamma}_2 = 0.64$.

In Fig. 5.8c, we demonstrate the SNR wall for pLS diversity under fading. We choose $L_1 = 0.5\,dB$ and $L_2 = 0.3\,dB$, and hence using Eq. (5.45), we get the threshold as $\lambda = 1.122$. Using Eq. (5.41) and Eq. (5.74), we get the SNR walls as $\bar{\gamma}_1 = 1.82$ and $\bar{\gamma}_2 = 1.71$. As discussed in Section 5.4.3, any one diversity branch should have SNR $\geq$ its respective SNR wall value to achieve unlimited reliability. In Fig. 5.8c, we choose $\bar{\gamma}_1 = 0.1$ which is below to its SNR wall value and $\bar{\gamma}_2 = 1.71$ which is equal the SNR wall. The vertical line in Fig. 5.8c shows that unlimited performance can be obtained by using the given setting. As done in case of pLC, one can consider different scenarios to obtain SNR wall in this case as well.

We next validate the SNR wall for hard combining obtained in Section 5.4.4 and listed in TABLE 5.2. In Fig. 5.9a, we show the plot of $\tau$ VS. $\bar{Q}_F$ and $\bar{Q}_D$, for $k$ out of $M$ rule considering $L_1 = 1\,dB$, $L_2 = 0.7\,dB$, $L_3 = 0.5\,dB$, $k = 2$ and $M = 3$. For these parameters, we obtained the SNR walls as listed in TABLE 5.2. In Fig. 5.9a, we show the plots by choosing $\bar{\gamma}_1 = 0.3$ which is below the required value of SNR wall and $\bar{\gamma}_2 = 2.62$ and $\bar{\gamma}_1 = 2.49$ which are equal to their respective SNR walls. Since

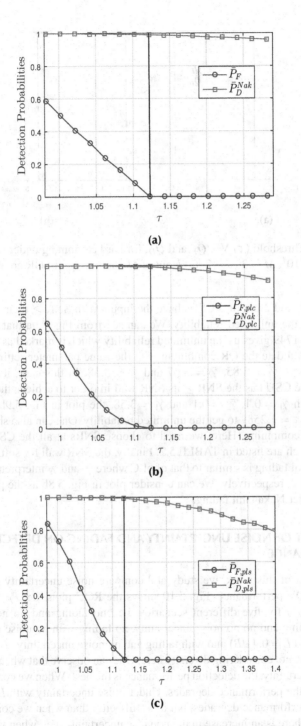

**Figure 5.8** Threshold ($\tau$) VS. detection probabilities for no diversity under Nakagami fading channel using $p = 2$, $N = 10^7$ for (a) no diversity with $L = 0.5$ dB, $\bar{\gamma} = 1.82$, (b) pLC diversity using $L_1 = L_2 = 0.5$ dB, $\bar{\gamma}_1 = \bar{\gamma}_2 = 0.67$, and (c) pLS diversity using $L_1 = 0.5$ dB, $L_2 = 0.3$, $\bar{\gamma}_1 = 0.1$, $\bar{\gamma}_2 = 1.71$.

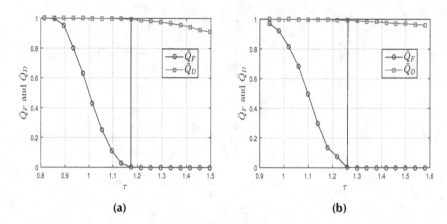

**Figure 5.9** Threshold ($\tau$) VS. $\bar{Q}_F$ and $\bar{Q}_D$, for hard combining under fading channel using $N = 10^7$, $M = 3$, $p = 2$, (a) $k$ out of $M$ combining, (b) OR combining.

we have $k = 2$, and 2 out of 3 CSUs have the inputs with SNR $\geq$ their SNR walls, we should get the unlimited reliability. We can see from Fig. 5.9a that choosing a value of $\tau = 1.1749$ gives us the unlimited reliability which is marked as the vertical line. We next validate the OR combining with the same parameter settings. We get the SNR walls as $\bar{\gamma}_1 = 3.85$, $\bar{\gamma}_2 = 3.67$ and $\bar{\gamma}_1 = 3.58$. In this case, it is required that at least one CSU has the SNR $\geq$ its SNR wall in order to achieve the unlimited reliability. With $\bar{\gamma}_1 = 0.1$, $\bar{\gamma}_2 = 0.1$ and $\bar{\gamma}_1 = 3.58$, the plot in Fig. 5.9b shows the vertical line at $\tau = 1.2589$ to get the unlimited reliability. One can also show similar plot for AND combining. Here, we need to choose SNRs at all the CSUs $\geq$ their SNR walls which are listed in TABLE 5.2. Finally, the SNR wall for soft combining under Nakagami fading is similar to that of pLC where $\bar{\gamma}_1$ and $\bar{\gamma}_2$ interpreted as SNRs at CSUs 1 and 2, respectively. We can consider plot in Fig. 5.8b as the plot for soft combining under Nakagami fading.

### 5.5.3 EFFECT OF NOISE UNCERTAINTY AND FADING ON DETECTION PERFORMANCE

As a byproduct of this work, we study and compare noise uncertainty effects and fading on GED's performance. Fig. 5.10 shows the ROC plots, i.e, $\bar{P}_F$ Vs. $\bar{P}_D$, for no diversity case for five different scenarios, i.e., no fading and no noise uncertainty, with fading and no no noise uncertainty, no fading with no noise uncertainty ($L = 0.2\ dB$ and $L = 0.4\ dB$) and with fading and no noise uncertainty ($L = 0.2\ dB$). Note that we are not considering diversity here. One can observe that when there is no fading and uncertainty, the detection performance is the best. When we consider Nakagami fading, the performance degrades. Under noise uncertainty with $L = 0.2\ dB$, the detection performance degrades, but it is still better than when we consider only fading. When there is an increase in the no noise uncertainty, i.e., when we consider

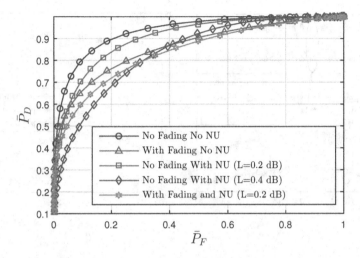

**Figure 5.10** $\bar{P}_F$ Vs. $\bar{P}_D$ considering different scenarios, i.e., no fading and no noise uncertainty, with fading and no noise uncertainty, no fading with noise uncertainty ($L = 0.2\ dB$ and $L = 0.4$), and with fading and noise uncertainty ($L = 0.2\ dB$). Here, $N = 500$, $\bar{\gamma} = -10\ dB$, $p = 2$, $m = 2$.

$L = 0.4\ dB$, we observe a significant degradation in detection performance, i.e., it becomes worse than the only fading case. For example, for $\bar{P}_F \approx 0.1$ in Fig. 5.10, the $\bar{P}_D$ with $L = 0.4\ dB$ is 20.8946% smaller than when we consider only fading. Often researchers working in the area of SS [60, 105, 199] assume that the true value of the noise variance is known, and hence the uncertainty in the noise variance is ignored. However, we observe that for certain noise uncertainty value, its effect is as severe as fading. Also, above some uncertainty level, the effect is more severe than fading, for example, in Fig. 5.10, when only uncertainty is considered with $L = 0.4\ dB$, the performance is well below that of fading, which is clearly evident in the ROC plots indicating degradation in the performance. Note that this happens for all values of $\bar{P}_F < 0.4$ in Fig. 5.10. We have also shown ROC plots when both the fading and no noise uncertainty exist by considering $L = 0.2\ dB$. We see that the performance is poor in this case compared to no fading no noise uncertainty, only fading and only noise uncertainty with $L = 0.2\ dB\ dB$. Nevertheless, what we observe is that the performance when only uncertainty is considered with $L = 0.4\ dB$ becomes worst for values of $\bar{P}_F$ below 0.3. Thus the plots show the severity of the effect of no noise uncertainty. Note that, although we have given plots considering no diversity only, one can obtain similar plots considering diversity. The conclusions drawn here for no diversity cases are also valid for diversity cases.

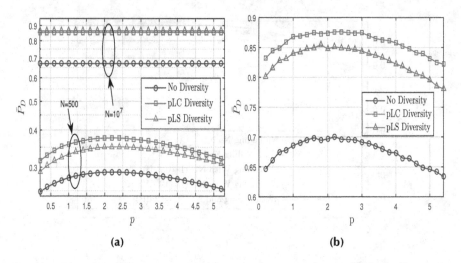

**Figure 5.11** $p$ Vs. $\bar{P}_D$ for no diversity, pLC diversity and pLS diversity for $\bar{P}_F = 0.1$ under fading and noise uncertainty considering (a) $L_1 = L_2 = L = 0.1 \, dB$, $m = 2$, $\bar{\gamma}_1 = \bar{\gamma}_2 = \bar{\gamma} = -15 \, dB$ with $N = 500$ and $N = 10^7$ and (b) $L_1 = L_2 = L = 0.5 \, dB$, $m = 2$, $\bar{\gamma}_1 = \bar{\gamma}_2 = \bar{\gamma} = -5 \, dB$ with $N = 100$.

### 5.5.4 EFFECT OF $p$

Finally, in Fig. 5.11, we show the plots of $p$ Vs. $\bar{P}_D$ in the presence of fading and no noise uncertainty considering no diversity, pLC, and pLS diversity by fixing $\bar{P}_F$ for different values of $N$ and noise uncertainties. We consider two diversity branches, i.e., $M = 2$, when we consider diversity. Since it is desired to have significantly less $\bar{P}_F$ in CR (e.g., in IEEE 802.22, we require $\bar{P}_F \leq 0.1$ [177]), we set $\bar{P}_F = 0.1$ and compute the threshold to determine $\bar{P}_D$. The authors in [24] show that the performance of GED is better for values of $p$ other than $p = 2$ where they do not consider noise uncertainty. However, they consider a small number of samples for detection. We know that the detection probability is directly proportional to the sample size $N$. One has to choose sufficiently high $N$ to achieve a higher probability of detection. Considering this, our plots here are obtained by taking $N = 500$, which is relatively high and $N = 10^7$, which is used to approximate an infinite sample size. It can be observed from Fig. 5.11a that for $N = 500$, the detection performance is best when the value of $p$ is close to 2 and for other values the probability of detection deteriorates. It can also be seen that with the use of diversity, the detection performance improves significantly. For example, with $p = 2$ and $N = 500$, we get 30.2% higher $\bar{P}_D$ using pLC diversity when compared to no diversity. As expected, pLC diversity performs better than pLS [164]. For example, the pLC diversity achieves 7.2244% higher probability of detection than pLS diversity for $p = 2$. Looking at the plots in Fig. 5.11 for $N = 10^7$, one may easily verify that as $N \to \infty$, the effect of $p$ vanishes. For example, for $N = 10^7$, the probability of detection is the same for any value of $p$

and the performance is almost the same for both pLC and pLS diversity. Fig. 5.11b shows the plots for $p$ Vs. $\bar{P}_D$ for $N = 100, L_1 = L_2 = 0.5\ dB$ and $\bar{\gamma}_1 = \bar{\gamma}_2 = -5\ dB$. It can be observed that, the GED with $p$ close to 2 performs better even for other values of noise uncertainties, SNRs, and the number of collected samples detection.

## 5.6 CONCLUSION

In this chapter, we studied the SNR wall for a generalized energy detector under no diversity, diversity, and CSS in the presence of both the noise uncertainty and the fading. We consider hard as well as soft combining for CSS. We derive closed form expressions for $\bar{P}_F$ and $\bar{P}_D$ for no diversity, diversity, and CSS considering a very large sample size. Using those expressions, we first derive the SNR wall for the AWGN channel case under noise uncertainty. This analysis is then extended to the case when there exists Nakagami fading in addition to noise uncertainty in which the SNR wall is obtained numerically. All the derived expressions are validated using Monte Carlo simulations. As the byproduct of this work, we also discussed and compared the effects of noise uncertainty and fading on GED performance. It is shown that above a certain value, noise uncertainty is more severe when compared to fading. It is also shown that for the value of $p$ close to 2, the GED performs the best, and the pLC diversity scheme outperforms the pLS scheme. For a large value of sample size, the detection performance of GED becomes independent of $p$, and the performance is the same for both the diversity schemes. As a future work, one may consider an analysis of the SNR wall for GED by considering noise uncertainty and more general fading models.

# Part II

## Wideband Spectrum Sensing

# 6 Diversity for Wideband Spectrum Sensing under Fading

In all our previous works, we looked at the problem of narrowband spectrum sensing. In this chapter, we consider wideband spectrum sensing (WSS), where multiple frequency bands are sensed for finding the spectrum opportunities. If the SUs can sense multiple bands at a time, they provide multiple spectrum opportunities. Authors in [154, 155], propose the optimal multiband joint detection for WSS in CR. The bank of multiple narrowband detectors is jointly optimized to improve the aggregate opportunistic throughput of a CR system while limiting the PU system's interference. The authors also propose a cooperative WSS scheme that exploits the spatial diversity among multiple CRs to improve sensing reliability. The Nyquist sampling rate required to sample the wideband is very high and can be practically challenging. To overcome this problem, many researchers have attempted the use of compressed sensing to acquire the signal at sub-Nyquist rates [14, 51, 134, 183, 194, 219]. However, these sampling schemes are complex, and their implementation is expensive. Many a times, a SU may not be interested in finding all the spectrum opportunities, instead, the interest lies in finding sufficient spectrum opportunities. As an example, a SU may be interested in finding a single spectrum opportunity for its transmission. This can be achieved if we consider only a part of the wideband spectrum for sensing. Taking this into consideration, authors in [185] propose partial band Nyquist sampling (PBNS), which samples part of the wideband instead of entire wideband, thus reducing the sampling rate. Since PBNS uses traditional Nyquist sampling, it represents the simplest WSS scheme.

The researchers have proposed different detection techniques for finding the presence or absence of signal while using WSS. Authors in [185], have proposed two detection algorithms, namely, channel-by-channel detection (CCD) and ranked channel detection (RCD), by considering the primary signal and the noise as additive white Gaussian noise. But, in practice, the received signal often undergoes fading. Limited work is done on WSS considering the fading [56, 181]. In [33], the performance of these detection algorithms is analyzed for two different fading channels, namely, Rayleigh and Nakagami fading. None of these approaches consider diversity reception. It may be of interest to note that the performance can be improved by using diversity reception. Little work is done on the use of diversity to improve the performance of WSS [12, 160]. Here in this work, we propose three new detection algorithms: channel-by-channel square law combining (CC-SLC), ranked square law combining (R-SLC) and ranked square law selection (R-SLS), where SLC and SLS diversities are used to improve the performance of WSS under fading. In our work, we use PBNS as the wideband sampling scheme. We provide a complete theoretical analysis of the proposed detection algorithms under the Nakagami fading channel. Nakagami fading is a generalized fading model and includes Rayleigh, Rice, and Hoyt fading models as its special cases. It can be used to model propagation in urban and sub-urban areas by setting the Nakagami parameter $m_1$. For quantitative analysis, we use recently proposed performance metrics, namely, the probability of excessive interference opportunity ($P_{EIO}$) and the probability of insufficient spectrum opportunity ($P_{ISO}$) [185]. Mathematical analysis is verified by Monte Carlo simulation. We also study the effect of different parameters on the performance of the proposed algorithms. We observe that proposed algorithms outperform RCD without diversity. The algorithm R-SLC and R-SLS outperforms CC-SLC algorithm. Also, the R-SLC algorithm performs better than R-SLS. Note that throughout this chapter, when detection algorithm is used without the diversity it is referred to as RCD and when used with the diversity they are referred to as R-SLC and R-SLS.

Before we proceed to a detailed discussion of our work, we list the crucial notations used in TABLE 6.1.

## 6.1 SYSTEM MODEL AND PERFORMANCE METRICS

The wideband signal is modeled as the collection of $U$ subbands, i.e., narrowband channels, each of bandwidth $B_0$ making the total bandwidth as $B = UB_0$. In [14, 15, 86, 152, 184, 185], similar channel model is used. The primary transmission within each subband is subject to flat fading. Let $H_m$ denotes the primary occupancy of the $m^{th}$ subband, with $H_m = 0$ and $H_m = 1$ corresponding to PU being OFF and ON, respectively. We assume that the occupancy status of the PU remains unchanged during the observation interval. This assumption is valid as long as the observation is not too long to become comparable to the primary ON/OFF time. It is also assumed that all the PUs have equal priorities. In this work, the discussion is restricted to one sensing window with a fixed duration of $T_w = UNT_N$, where $F_N = 1/T_N = 2B$ is the Nyquist sampling frequency, and $N$ represents a number of Nyquist samples per subband. We want to mention here that the terms subband and channel are used interchangeably.

We assume that the occupancy status of subbands and their transmissions are independent of each other. Assuming the occupancy probability (OP) of subbands as $p$, the state of each subband is modeled by a Bernoulli random variable and the probability mass function (PMF) of the same given as

$$f(H_m, p) = \begin{cases} p, & \text{if } H_m = 1 \\ 1-p, & \text{if } H_m = 0. \end{cases} \tag{6.1}$$

With the number of ON channels as $K$, PMF of primary occupancy is given by Binomial distribution as

$$f(K, U, p) = \binom{U}{K} p^K (1-p)^{U-K}. \tag{6.2}$$

**Table 6.1**
**Table of Notations**

| Symbols | Meaning |
|---|---|
| $U$ | Number of subbands in the wideband. |
| $B_0$ | Bandwidth of a subband. |
| $B$ | Bandwidth of wideband signal, i.e., $B = MB_0$. |
| $N$ | Nyquist samples per subband. |
| $F_N$ | Nyquist sampling frequency, i.e., $F_N = 2B$. |
| $T_N$ | Nyquist sampling duration, i.e., $T_N = 1/F_N$. |
| $T_w$ | Sensing window of duration $T_w = MNT_N$ in time. |
| $p$ | Occupancy probability of a PU subband. |
| $K$ | Number of ON channels in the wideband. |
| $y_m(t)$ | Received signal at SU due to $m^{th}$ primary subband. |
| $s_m(t)$ | Transmitted signal in the $m^{th}$ primary subband. |
| $n_m(t)$ | AWGN with mean 0 and variance $\sigma^2$ at the SU. |
| $h_m$ | Channel coefficient for the $m^{th}$ primary subband. |
| $\tau$ | Detection threshold. |
| $S$ | Number of spectrum opportunity, i.e., number of successfully identified OFF channels. |
| $S_d$ | Desired number of spectrum opportunity |
| $I$ | Number of missed ON channels. |
| $I_d$ | Maximum number of PU channels on which the interference from SU is allowed. |
| $P$ | Number of diversity branches. |
| $L$ | Number of subbands in the partial band. |
| $L_d$ | Number of subbands on which the decision is made. |
| $m_1$ | Nakagami parameter. |
| $\bar{\gamma}$ | Average signal to noise ratio. |
| $_1F_1(\cdot;\cdot;\cdot)$ | Confluent hypergeometric function [4, 13.1.2]. |
| $\chi^2_N$ | Central chi-square distribution with degree of freedom (DOF) $N$. |
| $1(t)$ | Unit step function. |
| $\Gamma(x)$ | Gamma function [4, 6.1.1]. |
| $\gamma(a,x)$ | Lower incomplete Gamma function [4, 6.5.2]. |
| $\Gamma(a,x)$ | Upper incomplete Gamma function [4, 6.5.3]. |

Note that to detect the free primary channels, the detector has to decide on their presence or absence. The received signal at the SU due to $m^{th}$ primary subband can now be modeled as

$$y_m(t) = h_m \cdot s_m(t) + n_m(t), \qquad (6.3)$$

where, $s_m(t)$ represents the transmitted signal in the $m^{th}$ primary subband, $n_m(t)$ represents AWGN and $h_m$ represents the channel coefficient for the $m^{th}$ primary subband. Note that, in this chapter we do not consider noise uncertainty.

Authors in [81, 86, 134, 194, 210] use mean-square estimation error (MSE) as the criteria for measuring performance. In [14, 84, 152, 183], the probability of miss detection ($P_M$) and false alarm ($P_F$) or their average over the channels are used as performance measures. These measures are more suitable for single channel detection. However, the goal of WSS can be different from that of single channel sensing. The SUs in CR may not be interested in finding all the spectrum opportunities. A fraction of spectrum opportunities may be sufficient for the SUs, and hence WSS can tolerate much higher $P_F$. Keeping this in perspective, new performance metrics are proposed in [185] that are more suitable for WSS. We use these two new performance metrics, i.e., $P_{EIO}$ and $P_{ISO}$, to characterize the performance of the detection algorithms. Let $S$ be the number of spectrum opportunity, i.e., the number of successfully identified OFF channels and let $S_d$ be the desired number of spectrum opportunity. Then $P_{ISO}$ is defined as

$$P_{ISO} = Pr\{S < S_d\}, \qquad (6.4)$$

where, $Pr\{\cdot\}$ represents probability. Similarly, $P_{ISO}$ is defined as

$$P_{EIO} = Pr\{I > I_d\}. \qquad (6.5)$$

where, $I$ represents the number of missed ON channels, and $I_d$ corresponds to the maximum number of allowed interference to the primary channels.

## 6.2 DETECTION ALGORITHMS

Various diversity schemes have been studied in the literature [19, 59, 60, 90, 170]. To name a few, maximum ratio combining (MRC), equal gain combining (EGC), square law selection (SLS), switched combining, square law combining (SLC), etc. MRC represents the optimal combining technique, and it gives the best possible performance that can be achieved by using diversity. However, the disadvantage here is that it requires complete channel state information (CSI) at the SU, and hence the complexity is very high. EGC is a suboptimal diversity technique and has reduced complexity compared to the MRC technique. Although it does not require channel fading amplitudes, we still need to use channel carrier phase estimation. The SLS combining technique selects the strongest signal branch for detection and is simple to implement compared to MRC and EGC. However, it requires continuous monitoring of all the diversity branches. The performance of switched combining diversity is inferior to SLS [110]. The SLC is a non-coherent diversity combining technique and is easy to implement. It performs better than SLS diversity [164]. In the SLC diversity scheme, the decision statistic obtained at all the diversity branches is added to obtain a new decision statistic, and the decision is taken based on that. Both SLS and SLC are non-coherent, combining schemes and are easy to implement. Considering all the above points, three new detection algorithms are proposed: channel-by-channel square law combining (CC-SLC), ranked square law combining (R-SLC) and ranked square law selection (R-SLS).

### 6.2.1 CHANNEL-BY-CHANNEL SQUARE LAW COMBINING (CC-SLC)

Consider $P$ diversity branches that receive a wideband signal consisting of $U$ channels. The PBNS is used at the input of all the diversity branches. As already discussed earlier, a SU may not be interested in finding all the spectrum opportunities; instead, the interest lies in finding sufficient numbers of spectrum opportunities. This motivates the idea of PBNS, which samples only the fraction of the entire wideband at the corresponding Nyquist rate. The PBNS is characterized by the number of channels in the partial band ($L$) and filters out $L$ channels from the wideband with $M$ channels. After filtering, the received signal is sampled at the Nyquist rate of $2LB_0$ on which the fast Fourier transform (FFT) is performed. We take $N$ samples for each channel. Hence, the number of samples in the sensing window is $NL$. Let $V[k]$ denote the frequency samples in PBNS. Then, $V[k] = Y[k]$, for $0 \le k \le NL - 1$, where $Y[k]$ denotes the normalized discrete Fourier transform (DFT) of the received wideband signal consisting of $U$ channels. The energy within $m^{th}$ narrowband channel is then calculated as

$$T_m = \sum_{k \in I_m} |V[k]|^2, \qquad (6.6)$$

where, $1 \le m \le L$ and $I_m$ is the set of frequency indices that fall into channel $m$.

Fig. 6.1, represents the block schematic of the proposed CC-SLC detection scheme, and the steps involved in implementing CC-SLC are given in Algorithm 1. As shown in Fig. 6.1, energies are computed for each of the $L$ channels from all $P$ diversity branches. The energies within each of the $L$ channels from all the diversity branches are added to obtain $T_1, T_2, \ldots, T_L$. The decision statistic obtained for $L$ channels is compared with the threshold $\tau$ to decide each channel's occupancy status. In Fig. (6.1), channel number 1 and $L$ are shown as free, and channel number 2 is shown as occupied. Note that one can use the square law selection diversity technique instead of square law combining technique and design channel-by-channel square law selection (CC-SLS) algorithm. Since the SLC diversity technique performs better than SLS diversity, we choose not to give CC-SLS algorithm here.

**Algorithm 1** Channel-by-Channel Square Law Combining (CC-SLC)

1: Compute energy for each of the $L$ channels from all $P$ diversity branches.
2: Add the energy within each of the $L$ channels from all the diversity branches. The resulting new energies are denoted by $T_m$, $m = 1, 2, \ldots, L$.
3: Make decisions on $L$ channels using the energy detection method based on the given threshold $\tau$.

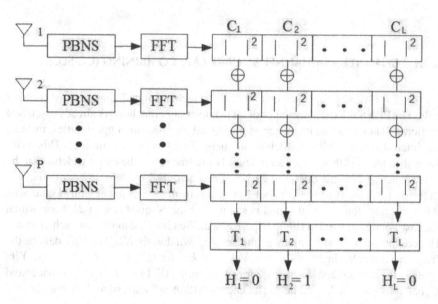

**Figure 6.1** Block diagram of channel-by-channel square law combining (CC-SLC) detection.

### 6.2.2 RANKED SQUARE LAW COMBINING (R-SLC) DETECTION

Figure 6.2, represents the block schematic of R-SLC detection schemes, and the steps involved in implementing it are given in Algorithm 2. As shown in Fig. 6.2, energies are computed for each of the $L$ channels from all $P$ diversity branches. The energies within each of the $L$ channels from all the diversity branches are added to obtain $T_1, T_2, \ldots, T_L$. Once this is done, the channels are ranked, i.e., arranged in ascending order with regard to $T_m$, $m = 1, 2, \ldots, L$. Then the decision is taken on first $L_d$ channels, where $L_d \leq S_d \leq L$. Remaining $U - L_d$ channels are ignored. Referring to Fig. 6.2, let $L_d = 2$ and let channels $L$ and 2 have the least received energy. Based on the decision made, channels $L$ and 2 are declared free and occupied, respectively. Note that, considering $P = 1$ in R-SLC results in RCD proposed in [185]. The main difference between CC-SLC and R-SLC is that in CC-SLC added energies are not arranged in ascending order before making the decision.

## Algorithm 2 Ranked Square Law Combining (R-SLC)

1: Compute energy for each of the $L$ channels from all $P$ diversity branches.
2: Add the energy within each of the $L$ channels from all the diversity branches. The resulting new energies are denoted by $T_m$, $m = 1, 2, \ldots, L$.
3: Sort the selected channels in ascending order with regard to $T_m$.
4: Make decisions for first $L_d$ channels, where $S_d \leq L_d \leq L$, using energy detection with the given threshold $\tau$. Ignore the remaining $U - L_d$ channels.

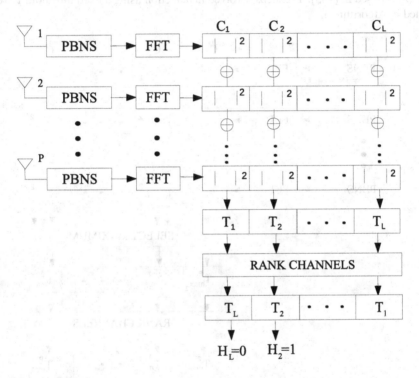

**Figure 6.2** Block diagram of ranked square law combining (R-SLC) detection.

### 6.2.3 RANKED SQUARE LAW SELECTION (R-SLS) DETECTION

Fig. 6.3 represents the block diagram of R-SLS detection scheme. As shown in Fig. 6.3, energy computed within each of the $L$ channels from all the diversity branches are given as input to a selector. The selector selects highest energy for each of the $L$ channels. Once, this is done, the selected channels are ranked, i.e., arranged in ascending order with regard to $T_m$, $m = 1, 2, \ldots, L$. Then the decision is taken on first $L_d$ channels, where $L_d \leq S_d \leq L$. Remaining $U - L_d$ channels are ignored. Fig. 6.3 represents the case with $L_d = 2$. Here, $C_1, C_2, \ldots, C_L$ represent $L$ narrowband channels. Here, diversity branches $P$ and 1 are shown to be selected for channel 1 and 2, i.e., $T^P_{max_1}$ and $T^1_{max_2}$, respectively. As shown in Fig. 6.3, diversity branches $P$ and 1 have maximum energies $T^P_{max_1}$ and $T^1_{max_2}$ for channels $C_1$ and $C_2$, respectively. Note that, in $T^j_{max_i}$, $max_i$ and $j$ represent the selected maximum energy in channel i and the corresponding diversity branch j, respectively, where $i = 1, 2, \ldots, L$ and $j = 1, 2, \ldots, P$. After ranking $C_L$ is shown to have the lowest energy and $C_2$ has next lowest energy from the selected maximum energies. The detection is made on these channels $L$ and 2. Based on threshold $\tau$, $C_L$ and $C_2$ are shown as free and occupied, respectively. Note that reducing the number of diversity branches to one results in RCD proposed in [185]. The steps involved in detection using a fixed threshold $\tau$ are listed in Algorithm 3.

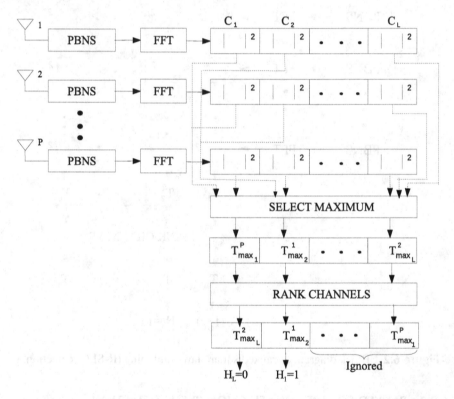

**Figure 6.3** Block diagram of ranked square law selection (R-SLS) detection.

---

**Algorithm 3** Ranked Square Law Selection (R-SLS)

1: Select maximum energy for each of the $L$ channels from all $P$ diversity branches.
2: Sort the selected channels in ascending order with regard to $T_m$.
3: Make decisions for first $L_d$ channels, where $S_d \leq L_d \leq L$, using energy detection with the given threshold $\tau$. Ignore the remaining $U - L_d$ channels.

## 6.3 APPROXIMATION OF DECISION STATISTIC

The probability density function (pdf) of received energy under Nakagami fading is given as [105],

$$f_{T_m}(t) = \begin{cases} \frac{1(t)t^{G-1}e^{-\frac{t}{2}}}{(G-1)!2^G D} {}_1F_1(m_1; G; tE), & H_m = 1 \\ \chi_N^2, & H_m = 0 \end{cases} \quad (6.7)$$

where, $\chi_N^2$ is the chi-square pdf with $N$ degrees of freedom, $D = (1+\bar{\gamma}/m_1)^{m_1}$, $E = 0.5 - 0.5m_1/(m_1+\bar{\gamma})$, $\bar{\gamma}$ is the average SNR, $G = N/2$, $M = G-1$, $m_1$ is the Nakagami parameter, ${}_1F_1(\cdot;\cdot;\cdot)$ is the confluent hypergeometric function [79] and $1(t)$ is the unit step function.

If the pdf under $H_m = 1$ is directly used to obtain the expressions for performance metrics, they lead to infinite series representation and one may find it difficult to draw insight of different parameters from these expressions. In order to resolve this problem, we first approximate the pdf of $T_m$ under $H_m = 1$ using a simple expression by performing the asymptotic analysis. Using the Taylor series expansion of a function $f(t)$, we can write [58]

$$f(t) = at^q + a_1 t^{q+1} + O(t^{q+2}) \text{ as } t \to 0^+. \quad (6.8)$$

Here, $a$, $a_1$ and $q$ represent real constants and $O(t^{q+2})$ is the error term as $t \to 0^+$. Using Eq. (6.8), the authors in [58] propose the approximation as

$$f(t) \approx at^q e^{-\alpha t}, \text{ as } t \to 0^+, \quad (6.9)$$

where, $\alpha = -\frac{a_1}{a}$. The Taylor series expansion of Eq. (6.7) under $H_m = 1$ can be obtained as

$$f_{T_m}(t) = \frac{1}{2^G D \Gamma(G)} t^{G-1} + \frac{2m_1 E - G}{2^{G+1} D \Gamma(G+1)} t^G + O(t^{G+1}), \quad (6.10)$$

as $t \to 0^+$. On comparing Eq. (6.10) with Eq. (6.8) we get $a = \frac{1}{2^G D \Gamma(G)}$, $a_1 = \frac{(2m_1 E - G)}{2^{G+1} D \Gamma(G+1)}$, $\alpha = -\frac{a_1}{a}$ and $q = G - 1$. With this, the approximated pdf of $T_m$ is given by,

$$f_{T_m}(t) = \begin{cases} 1(t) a e^{-\alpha t} t^q, & H_m = 1 \\ \chi_N^2. & H_m = 0 \end{cases} \quad (6.11)$$

This approximation may not result in proper pdf, i.e., the area under $f_{T_m}(t)$ is not necessarily 1. However, it better approximates expression given in Eq. (6.7) under low $\bar{\gamma}$. In Fig. 6.4, we show the plots for actual and approximated pdf for different values of $\bar{\gamma}$. The plots are shown for $N = 10$, $m_1 = 2$ for $\bar{\gamma} = 0$ dB, $\bar{\gamma} = -5$ dB and $\bar{\gamma} = -10$ dB. One can see that for $\bar{\gamma} = 0$ dB, the approximation slightly deviates from the actual pdf, whereas for $\bar{\gamma} = -5$ dB and $\bar{\gamma} = -10$ dB the approximated pdf almost overlaps with the actual pdf. For $\bar{\gamma} > 0\ dB$, we observe that the approximation deviates from the actual pdf. Hence, we conclude that the approximated pdf can be used to analyze the performance of the algorithms for $\bar{\gamma} \leq 0\ dB$. This is acceptable since in order to implement the CR without interference to the PU, it is important to be able to detect the existence of the PU under very low SNR environment. Also, if the detection algorithms perform better under low SNR, they also perform well under high SNR since the density functions under $H_0$ and $H_1$ are well separated.

### 6.3.1 PDF FOR SLC DIVERSITY

As already discussed, under the SLC diversity scheme, the measured energy from different diversity branches are added to obtain the new decision statistic [59, 60]. Assuming same average SNR ($\bar{\gamma}$) for all the diversity branches, pdf of sums of independent random variables can be obtained by convolving their marginal pdfs. Considering received energy as independent and identically distributed (iid) random variable, the pdf of decision statistic for SLC can be obtained by convolving branch pdfs of $T_m$, $m = 1, 2, \cdots, L$. To understand the need for approximation, the pdfs without and with using approximations are derived.

#### 6.3.1.1 Without Using Approximation

It is mathematically involved to derive the decision statistic using SLC diversity considering any number of diversity branches. Hence, here we derive the pdf using two diversity branch only. The new decision statistics considering Nakagami fading under SLC diversity is derived by convolving two independent decision statistics under Nakagami fading, i.e., using Eq. (6.7). After convolving, we get the pdf of added energy $f_{T_m}(t)$ as

$$f_{T_m}(t) \sim \begin{cases} \sum_{v=0}^{\infty} \sum_{q=0}^{\infty} \frac{e^{-\frac{t}{2}}\Gamma(m_1+v)\Gamma(m_1+q)E^{v+q}t^{2G+v+q-1}}{2^{2G}D^2\Gamma(m_1)^2 v! q! \Gamma(2G+v+q)}, & H_m = 1 \\ \chi^2_{2N}. & H_m = 0 \end{cases} \quad (6.12)$$

The convergence of the infinite series under $H_m = 1$ in Eq. (6.12) can be verified using the comparison test and ratio test. We show that it indeed converges and the proof for the same is given in APPENDIX D.1.

#### 6.3.1.2 Using Approximation

Under $H_m = 0$, adding $P$ independent and identically distributed (iid) central chi-square random variables, each with $N$ degrees of freedom results in another chi-square random variable with $PN$ degrees of freedom [60]. The pdf of $T_m$ under $H_m = 1$ with $P$ diversity branches can be obtained by successive convolution of Eq. (6.11) with itself. With this, the pdf of decision statistic can be summarized as

$$f_{T_m}(t) \approx \begin{cases} \frac{1(t)a^P \Gamma(1+q)^P t^{Pq+P-1} e^{-\alpha t}}{\Gamma(Pq+P)}, & H_m = 1 \\ \chi^2_{PN}. & H_m = 0 \end{cases} \quad (6.13)$$

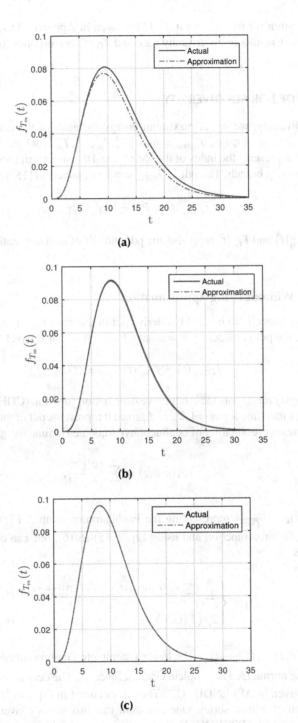

**Figure 6.4** Plots showing the actual and approximated pdf considering $N = 10$, $m_1 = 2$ for (a) $\bar{\gamma} = 0\ dB$, (b) $\bar{\gamma} = -5\ dB$, and (c) $\bar{\gamma} = -10\ dB$.

The derivation for the pdf in Eq. (6.13) is given in Appendix D.4. Note that substituting $P = 1$ results in no diversity case and $f_{T_m}(t)$ corresponds to that given in Eq. (6.11).

## 6.3.2 PDF FOR SLS DIVERSITY

In SLS diversity, we select maximum energy for each of the $L$ channels from all $P$ diversity branches, i.e., $T_{max_m} = \max\{T_m^1, T_m^2, \ldots, T_m^P\}$, where the subscript $m = 1, 2, \ldots, L$ represents the index of the channel and the superscript represents the index of the diversity branch. The pdf of $T_{max_m}$ can be obtained as [150]

$$f_{T_{max_m}}(t) = P(F_{T_m}(t))^{P-1} f_{T_m}(t), \quad (6.14)$$

where, $f_{T_m}(t)$ and $F_{T_m}(t)$ represent the pdf and cdf of decision statistic for $m^{th}$ channel.

### 6.3.2.1 Without Using Approximation

It is mathematically to involved to derive pdf in this case for any value of $P$. Hence, we derive the pdf considering $P = 2$ only. Using $P = 2$ in Eq. (6.14) we get

$$f_{T_{max_m}}(t) = f_{T_{max}}(t) = 2F_{T_m}(t)f_{T_m}(t), \quad (6.15)$$

where, $F_{T_m}(t)$ represents the cumulative distribution function (CDF) of $T_m$. Note that $f_{T_{max_m}}(t)$ is the same for $m = 1, 2, \ldots, L$, since it represents pdf of the received energy. Using series representation of confluent hypergeometric function given by

$$_1F_1(a,b,z) = \sum_{k=0}^{\infty} \frac{(a)_k}{(b)_k} \frac{z^k}{k!}, \quad (6.16)$$

where, $(a)_k = \frac{\Gamma(a+k)}{\Gamma(a)}$ representing the Pochhammer symbol, $\Gamma(z) = \int_0^\infty x^{z-1} e^{-x} dx$ being the Gamma function and using Eq. (6.15)-(6.16), one can obtain $f_{T_{max_m}}(t) = f_{T_{max}}(t)$ as

$$f_{T_{max}}(t) = \begin{cases} \sum_{k=0}^{\infty} \frac{t^{G-1} e^{-\frac{t}{2}} E^k (m_1)_k {}_1F_1(m_1, G, tE) \gamma(G+k, \frac{t}{2})}{(G)_k \Gamma(G)^2 2^{G-k-1} D^2 k!}, & H_m = 1 \\ 2\chi_N^2 P(G, \frac{t}{2}), & H_m = 0 \end{cases} \quad (6.17)$$

where, $\gamma(a,z) = \int_0^z \frac{x^{a-1}}{e^z} dx$ is the lower incomplete Gamma function and $P(a,z) = \frac{\gamma(a,z)}{\Gamma(a)}$ is the normalized incomplete Gamma function. The derivation of the pdf in Eq. (6.17) is given in APPENDIX D.2. The pdf obtained in Eq. (6.17) under $H_m = 1$ is in the form of infinite series. One can show that this series converges, the proof of which is given in APPENDIX D.3.

### 6.3.2.2 Using Approximation

Using $f_{T_m}(t)$ given in Eq. (6.11), one can derive $F_{T_m}(t)$ under $H_m = 1$ as

$$F_{T_m}(t) = a \int_0^t x^q e^{-\alpha x} dx. \qquad (6.18)$$

Using the change of variable as $y = \alpha x$, we get

$$F_{T_m}(t) = \frac{a}{\alpha^{q+1}} \int_0^{\alpha t} y^q e^{-y} dx. \qquad (6.19)$$

Using simple mathematical manipulations, the inner integral can be reduced to $\gamma(q+1, \alpha t)$. Using this, $F_{T_m}(t)$ under $H_m = 1$ can be written as

$$F_{T_m}(t) = \frac{a\gamma(q+1, \alpha t)}{\alpha^{q+1}}. \qquad (6.20)$$

Following a similar procedure, $F_{T_m}(t)$ under $H_m = 0$ can be obtained as

$$F_{T_m}(t) = \frac{\gamma(G, \frac{t}{2})}{\Gamma(G)}. \qquad (6.21)$$

Using these, we obtain the pdf of $T_{max_m}$ as

$$f_{T_{max_m}}(t) \approx \begin{cases} P\left[\frac{a\gamma(q+1,\alpha t)}{\alpha^{q+1}}\right]^{P-1} ae^{-\alpha t} t^q, & H_m = 1 \\ P\left[\frac{\gamma(G,\frac{t}{2})}{\Gamma(G)}\right]^{P-1} \frac{t^{G-1} e^{-\frac{t}{2}}}{2^G \Gamma(G)}, & H_m = 0 \end{cases} \qquad (6.22)$$

## 6.4 THEORETICAL ANALYSIS OF DETECTION ALGORITHMS

In this section, first we derive the expressions for $P_{ISO}$ and $P_{EIO}$ for CC-SLC and then the same are derived for R-SLC and R-SLS. Note that, in the subsequent explanation the superscript *Nak* represents the Nakagami fading.

### 6.4.1 CHANNEL-BY-CHANNEL SQUARE LAW COMBINING (CC-SLC)

In CC-SLC, the decision on each of the primary channels is made separately. The performance measures $P_{ISO}$ and $P_{EIO}$ are obtained by deriving the probability of miss detection ($P_m$) and the probability of false alarm ($P_f$) for each of these channels. These are related to probability of interference opportunity ($PI$) and the probability of spectrum opportunity ($PS$) as follows

$$PS(\tau) = (1-p)(1-P_f(\tau)), \qquad (6.23)$$

$$PI(\tau) = pP_m(\tau), \qquad (6.24)$$

where, $p$ represents occupancy probability of a channel. Using Eq. (6.23) and Eq. (6.24), $P_{ISO}$ and $P_{EIO}$ can be obtained as

$$P_{ISO}(S_d) = \sum_{k=0}^{S_d-1} \binom{L}{k} PS(\tau)^k (1-PS(\tau))^{L-k}, \qquad (6.25)$$

$$P_{EIO}(I_d) = \sum_{k=I_d+1}^{L} \binom{L}{k} PI(\tau)^k (1-PI(\tau))^{L-k}, \qquad (6.26)$$

respectively.

We derive $P_m$ and $P_f$ by considering SLC diversity with $P$ diversity branches using pdfs given in Eq. (6.13). The $P_{m,SLC}^{Nak}(\tau)$ for a particular threshold $\tau$ is defined as $P_{m,SLC}^{Nak}(\tau) = Pr\{T_m < \tau | H_m = 1\}$. It can be obtained by integrating the pdf in (6.13) under $H_m = 1$ from 0 to $\tau$ as

$$P_{m,SLC}^{Nak}(\tau) = \int_0^\tau \frac{a^P \Gamma(1+q)^P t^{Pq+P-1} e^{-\alpha t}}{\Gamma(Pq+P)} dt$$

$$= \frac{a^P \Gamma(1+q)^P}{\Gamma(Pq+P)} \int_0^\tau t^{Pq+P-1} e^{-\alpha t} dt. \qquad (6.27)$$

Applying change of variable as $y = \alpha t$, we get

$$P_{m,SLC}^{Nak}(\tau) = \frac{a^P \Gamma(1+q)^P}{\Gamma(Pq+P) \alpha^{Pq+P}} \int_0^\tau y^{Pq+P-1} e^{-y} dt. \qquad (6.28)$$

Note that the integral in Eq. (6.28) represents $\gamma(Pq+P, \alpha\tau)$. Using this, we can write $P_m^{Nak}(\tau)$ as

$$P_{m,SLC}^{Nak}(\tau) = \frac{a^P \Gamma(1+q)^P \gamma(Pq+P, \alpha\tau)}{\Gamma(Pq+P) \alpha^{Pq+P}}. \qquad (6.29)$$

The $P_f$ is defined as $P_f(\tau) = Pr\{T_m > \tau | H_m = 0\}$. This can be obtained by integrating pdf in Eq. (6.13) under $H_m = 0$ from $\tau$ to $\infty$ and is given as

$$P_{f,SLC}(\tau) = \frac{\Gamma(P \cdot G, \frac{\tau}{2})}{\Gamma(P \cdot G)}, \qquad (6.30)$$

where, $\Gamma(a,x) = \int_x^\infty t^{a-1} e^{-t} dt$ is the upper incomplete Gamma function. Here, dot $(\cdot)$ represents multiplication. Once $P_{mis,SLC}^{Nak}(\tau)$ and $P_{f,SLC}(\tau)$ are obtained, we can obtain $P_{ISO}(\tau)$ and $P_{EIO}(\tau)$ by using Eq. (6.23) to Eq. (6.26). Note that, if we use $P = 1$ in Eq. (6.29) and Eq. (6.30), it results in no diversity case.

## 6.4.2 THEORETICAL ANALYSIS FOR R-SLC

In this section, we derive $P_{ISO}$ and $P_{EIO}$ for ranked square law combining detection by considering Nakagami fading channel. To make the analysis easy to understand, we first consider only two channels in the partial band, i.e., $L = 2$ and then extend the analysis to the general case. We assume the desired number of spectrum opportunities as one, i.e., $S_d = 1$ and $I_d = 0$, i.e., no primary channel interference. We take a decision on one channel only, i.e., $L_d = 1$. We assume $S_d = 1$, $I_d = 0$ and $L_d = 1$. With these parameter settings, the energies for the two channels are computed which are represented as $T_1$ and $T_2$ for channel 1 and 2, respectively. These energies are ranked, i.e., they are arranged in ascending order by computing the minimum of the two energies, i.e., $T_{min} = min\{T_1, T_2\}$. Since $L_d = 1$, decision is made on $T_{min}$ and the other channels are declared ON. Now, if $T_{min} < \tau$, the channel is declared as OFF; otherwise, it is declared as ON. For example, suppose $T_2 < T_1$, then channel 1 is declared ON, and the decision is made on channel 2.

In order to obtain $P_{ISO}$ and $P_{EIO}$, we need to derive the probability of spectrum opportunity (*PS*) and the probability of interference opportunity (*PI*). Since we use $L = 2$, depending on the number of ON channels, there are three possible cases for primary occupancy of two channels, i.e., both channels are OFF, only one channel is ON, and both channels are ON.

If both the channels are OFF, the probability of spectrum opportunity, i.e., $PS_{00}^{Nak}(\tau)$ can be obtained by calculating probability of $T_{min} < \tau$ under $H_1 = H_2 = 0$. i.e.,

$$PS(\tau|H_1, H_2 = 0) = PS_{00}^{Nak}(\tau) = Pr\{T_{min} < \tau\}. \tag{6.31}$$

Here, $T_{min} = min(T_1, T_2)$. If we consider $L$ iid random variables, the pdf of minimum, i.e., $T_{min} = min(T_1, T_2, \ldots, T_L)$ is obtained as [150],

$$f_{T_{min}}(t) = Lf_T(t)(1 - F_T(t))^{L-1} = Lf_T(t)\bar{F}_T(t)^{L-1}, \tag{6.32}$$

where, $F_T(t)$ and $\bar{F}_T(t) = 1 - F_T(t)$ are the cumulative distribution functions (cdf) and the complementary cdf (ccdf) of $f_T(t)$, respectively.

To derive $f_{T_{min}}(t)$ in this case, we have to first derive $\bar{F}_T(t)$ which can be obtained by integrating $f_T(t)$ in Eq. (6.13) under $H_m = 0$ from $t$ to $\infty$. Using $\bar{F}_T(t)$ and Eq. (6.13) under $H_m = 0$ in Eq. (6.32) and substituting $L = 2$, one can obtain the pdf of $T_m$ when both the channels are OFF as

$$f_{T_{min}}(t) = \frac{t^{PG-1}e^{-\frac{t}{2}}\Gamma(P \cdot G, \frac{t}{2})}{2^{P \cdot G-1}\Gamma(P \cdot G)^2}. \tag{6.33}$$

Now, PS can be obtained by using series representation of $\Gamma(n,x)$ for integer $n$ from [13, 8.69] in Eq. (6.33) and integrating the same from 0 to $\tau$. After performing mathematical simplifications we obtain the expression for PS as

$$PS_{00}^{Nak}(\tau) = \sum_{k=0}^{P \cdot G-1} \frac{\gamma(P \cdot G + k, \tau)}{2^{PG+k-1} k! \, \Gamma(P \cdot G)}. \tag{6.34}$$

The derivation of Eq. (6.34) is given in APPENDIX D.5. Since both the channels are OFF, there is no chance of interference and hence $PI(\tau|H_1,H_2=0) = PI_{00}^{Nak}(\tau) = 0$.

We now consider the case when both the channels are ON. In this case, spectrum opportunity does not exist and hence $PS(\tau|H_1,H_2=1) = PS_{11}^{Nak}(\tau) = 0$. The probability of interference opportunity, i.e., $PI_{11}^{Nak}(\tau)$ in this scenario can be obtained by calculating the probability of $T_{min} < \tau$ under $H_1 = H_2 = 1$.

$$PI(\tau|H_1,H_2=1) = PI_{11}^{Nak}(\tau) = Pr\{T_{min} < \tau\}. \tag{6.35}$$

Once again using Eq. (6.13) but under $H_m = 1$ in Eq (6.32) and substituting $L = 2$, we obtain the pdf of $T_m$ as

$$f_{T_{min}}(t) = \frac{2a^P\Gamma(1+q)^P t^{Pq+P-1} e^{-\alpha t}}{\Gamma(Pq+P)} - \frac{2a^{2P}\Gamma(1+q)^{2P} t^{Pq+P-1} e^{-\alpha t} \gamma(Pq+P,\alpha t)}{\Gamma(Pq+P)^2 \alpha^{Pq+P}}. \tag{6.36}$$

The derivation of $f_{T_{min}}(t)$ in Eq. (6.36) is given in APPENDIX D.6. From this, $PI_{11}^{Nak}(\tau)$ is obtained by integrating Eq. (6.36) from 0 to $\tau$, which can be shown to be

$$PI_{11}^{Nak}(\tau) = \sum_{k=0}^{Pq+P-1} \frac{a^{2P}\Gamma(1+q)^{2P}\gamma(Pq+P+k,2\alpha\tau)}{\Gamma(Pq+P)\alpha^{2Pq+2P}k!2^{Pq+P+k-1}} + \frac{2a^P(q!)^P\gamma(Pq+P,\alpha\tau)\left(\alpha^{Pq+P} - a^P\Gamma(1+q)^P\right)}{\Gamma(Pq+P)\alpha^{2Pq+2P}}. \tag{6.37}$$

Finally, if only one channel is ON (assuming channel 1 is OFF and channel 2 is ON), $PS_{01}^{Nak}(\tau)$ can be obtained by calculating the probability of event $\{T_1 < T_2, T_1 < \tau\}$, i.e.,

$$PS(\tau|H_1=0, H_2=1) = PS_{01}^{Nak}(\tau) = Pr\{T_1 < T_2, T_1 < \tau\},$$
$$= \int_0^\tau \left(f_{T_1}(t_1) \int_{t_1}^\infty f_{T_2}(t_2) dt_2\right) dt_1,$$
$$= \int_0^\tau f_{T_1}(t_1) \bar{F}_{T_2}(t_1) dt_1, \tag{6.38}$$

where, $f_{T_1}(t)$ and $f_{T_2}(t)$ are the pdfs of energy received in channel 1 and 2, respectively, and $\bar{F}_{T_2}(t)$ is the ccdf of $f_{T_2}(t)$. Substituting pdfs from Eq. (6.13) into Eq. (6.38) and integrating using variable transformation, $PS_{01}^{Nak}(\tau)$ is given by

$$PS_{01}^{Nak}(\tau) = \sum_{k=0}^{Pq+P-1} \frac{a^P\Gamma(q+1)^P\gamma(P\cdot G+k,(\alpha+\frac{1}{2})\tau)}{2^{P\cdot G}\Gamma(P\cdot G)\alpha^{Pq+P-k}k!(\alpha+\frac{1}{2})^{P\cdot G+k}}. \tag{6.39}$$

The PI in this case can be obtained using

$$PI(\tau|H_1=0, H_2=1) = PI_{01}^{Nak}(\tau) = Pr\{T_2 < T_1, T_2 < \tau\},$$
$$= \int_0^\tau \left( f_{T_2}(t_2) \int_{t_1}^\infty f_{T_1}(t_1) dt_1 \right) dt_2,$$
$$= \int_0^\tau f_{T_2}(t_2) \bar{F}_{T_1}(t_2) dt_2. \quad (6.40)$$

The $PI_{01}^{Nak}(\tau)$ is derived by substituting pdfs from Eq. (6.13) into Eq. (6.40) and performing the integration. After simplification, it is given by

$$PI_{01}^{Nak}(\tau) = \sum_{k=0}^{P \cdot G - 1} \frac{a^P \Gamma(1+q)^P \gamma\left(Pq+P+k, \left(\alpha+\frac{1}{2}\right)\tau\right)}{\Gamma(Pq+P) 2^k k! \left(\alpha+\frac{1}{2}\right)^{Pq+P+k}}. \quad (6.41)$$

Note that, if we consider channel 1 as ON and channel 2 as OFF, we obtain the same results as in Eq. (6.39) and Eq. (6.41), respectively, i.e., $PS_{01}^{Nak}(\tau) = PS_{10}^{Nak}(\tau)$ and $PI_{01}^{Nak}(\tau) = PI_{10}^{Nak}(\tau)$. Once, PS and PI are available for all three cases, $P_{ISO}$ and $P_{EIO}$ can be obtained as

$$P_{ISO}^{Nak}(\tau) = 1 - \sum_{i=0}^{1} \sum_{j=0}^{1} (1-p)^{L-i-j} p^{i+j} PS_{ij}^{Nak}(\tau), \quad (6.42)$$

$$P_{EIO}^{Nak}(\tau) = \sum_{i=0}^{1} \sum_{j=0}^{1} (1-p)^{L-i-j} p^{i+j} PI_{ij}^{Nak}(\tau), \quad (6.43)$$

respectively.

The analysis considering 3 channels in the partial band is given in Appendix D.7. Looking at the analysis for $L=2$ and $L=3$, we observe that there exists a pattern and hence the analysis for the general case of any number of channels in the partial band ($L$) can be easily arrived at. To do this, let us use Eq. (6.13) to write the pdfs and ccdfs of the received decision statistic as follows

$$f_{H0}(t) = \frac{t^{P \cdot G - 1} e^{-\frac{t}{2}}}{2^{P \cdot G} \Gamma(P \cdot G)}, \quad \text{pdf under } H_m = 0, \quad (6.44)$$

$$f_{H1}(t) = \frac{a^P \Gamma(1+q)^P t^{Pq+P-1}}{e^{\alpha t} \Gamma(Pq+P)}, \quad \text{pdf under } H_m = 1, \quad (6.45)$$

$$\bar{F}_{H0}(t) = \frac{\Gamma(P \cdot G, \frac{t}{2})}{\Gamma(P \cdot G)}, \quad \text{ccdf of } f_{H0}(t), \quad (6.46)$$

$$\bar{F}_{H1}(t) = \frac{a^P \Gamma(1+q)^P \Gamma(Pq+p, \alpha t)}{\Gamma(Pq+P) \alpha^{Pq+P}}, \quad \text{ccdf of } f_{H1}(t). \quad (6.47)$$

Using these equations, $P_{ISO}^{Nak}(\tau)$ and $P_{EIO}^{Nak}(\tau)$ for any $L$ can be given by

$$P_{ISO}^{Nak}(\tau) = 1 - \sum_{i=0}^{L} \binom{L}{i}(1-p)^{L-i}p^i(L-i)\int_0^\tau f_{H0}(t)(\bar{F}_{H0}(t))^{L-i-1}(\bar{F}_{H1}(t))^i\,dt. \tag{6.48}$$

$$P_{EIO}^{Nak}(\tau) = \sum_{i=0}^{L} \binom{L}{i}(1-p)^{L-i}p^i\,i\int_0^\tau f_{H1}(t)(\bar{F}_{H1}(t))^{i-1}(\bar{F}_{H0}(t))^{(L-i)}\,dt. \tag{6.49}$$

Note that, if we use $P = 1$ in the analysis, it results in no diversity case, i.e., RCD. Also, Eq. (6.48) and Eq. (6.49) are general and can be used with any fading model once the pdf of decision statistic under that fading is available.

### 6.4.3 THEORETICAL ANALYSIS OF R-SLS

In this section we provide theoretical analysis of R-SLS detection algorithm. The analysis here is similar to that carried out for R-SLC in Section 6.4.2. We first consider a case with $L = P = 2$, $L_d = 1 = S_d = 1$, $I_d = 0$ and same SNRs at the input of all diversity branches and then extend it to general case.

Since we have $L = P = 2$, we first select maximum energy $T_{max_m}$ i.e., $T_{max_m} = max\{T_m^1, T_m^2\}$, $m = 1, 2$ represents the index of the channel and the superscripts 1 and 2 represent indices of the diversity branches. The pdf of $T_{max_m}$ is obtained by using $P = 2$ in Eq. (6.22). Now, there are three possible cases for occupancy of the two channels $C_1$ and $C_2$. These correspond to: 1. Both the channels are OFF, 2. Only one channel is ON and 3. Both the channels are ON. If both the channels are OFF, one can write the probability of spectrum opportunity (PS) as

$$PS(\tau|H_1 = 0, H_2 = 0) = PS_{00}(\tau) = 2\int_0^\tau \left(f_{T_{max_1}}(t_1)\int_{t_1}^\infty f_{T_{max_2}}(t_2)dt_2\right)dt_1, \tag{6.50}$$

where, $T_{min} = min\{T_{max_1}, T_{max_2}\}$. Here, $T_{max_1} = max\{T_1^1, T_1^2\}$ and $T_{max_2} = max\{T_2^1, T_2^2\}$. Since both the channels are OFF, we have $f_{T_{max_1}}(t_1) = f_{T_{max}}(t) = f_{T_{max}}(t)$ which is given in Eq. (6.22) under $H_m = 0$. It is mathematically too involved to arrive at the close form expression for Eq. (6.50) and hence we keep it in integral form only. Since both the channels are OFF, there is no chance of interference and hence, $PI(\tau|H_1 = 0, H_2 = 0) = PI_{00}(\tau) = 0$.

Next, we consider the case when only one channel is ON. Assuming channel 1 is OFF and channel 2 is ON, PS can be written as

$$PS(\tau|H_1 = 0, H_2 = 1) = PS_{01}(\tau) = Pr\{T_{max_1} < T_{max_2}, T_{max_1} < \tau\}. \tag{6.51}$$

Now, PS in this scenario can be obtained as

$$PS_{01}(\tau) = \int_0^\tau \left(f_{T_{max_1}}(t_1)\int_{t_1}^\infty f_{T_{max_2}}(t_2)dt_2\right)dt_1, \tag{6.52}$$

where, $f_{T_{max_1}}(t_1)$ and $f_{T_{max_2}}(t_2)$ are the pdfs given in Eq. (6.22) under $H_m = 0$ and $H_m = 1$, respectively. Similarly, the probability of interference opportunity (PI) can be written as

$$PI(\tau|H_1 = 0, H_2 = 1) = PI_{01}(\tau) = Pr\{T_{max_2} < T_{max_1}, T_{max_2} < \tau\}. \quad (6.53)$$

This can be obtained as

$$PI_{01}(\tau) = \int_0^\tau \left( f_{T_{max_2}}(t_2) \int_{t_2}^\infty f_{T_{max_1}}(t_1) dt_1 \right) dt_2. \quad (6.54)$$

Note that, one can obtain the same expressions for PS and PI assuming channel 1 as ON and channel 2 as OFF, i.e., $PS_{01}(\tau) = PS_{10}(\tau)$ and $PI_{01}(\tau) = PI_{10}(\tau)$.

Now, if both channels are ON, we have $PS(\tau|H_1 = 1, H_2 = 1) = PS_{11}(\tau) = 0$, as this results in zero spectrum opportunities. The PI in this case can be written as

$$PI(\tau|H_1 = 1, H_2 = 1) = PI_{11}(\tau) = 2 \int_0^\tau \left( f_{T_{max_1}}(t_1) \int_{t_1}^\infty f_{T_{max_2}}(t_2) dt_2 \right) dt_1. \quad (6.55)$$

Here, we consider both the channels as ON and hence $f_{T_{max_1}}(t_1) = f_{T_{max_2}}(t_2) = f_{T_{max}}(t)$ which is given by Eq. (6.22) under $H_m = 1$.

Using $PS_{ij}(\tau)$ and $PI_{ij}(\tau)$, for $i, j = 0, 1$, probability of a spectrum opportunity $(PS(\tau))$ and the probability of interference opportunity $(PI(\tau))$ for the selected threshold $\tau$ can be computed using

$$PS(\tau) = (1-p)^2 PS_{00}(\tau) + 2p(1-p) PS_{01}(\tau) + p^2 PS_{11}(\tau) \text{ and} \quad (6.56)$$

$$PI(\tau) = (1-p)^2 PI_{00}(\tau) + 2p(1-p) PI_{01}(\tau) + p^2 PI_{11}(\tau). \quad (6.57)$$

Finally, using Eq. (6.56) and Eq. (6.57), $P_{ISO}$ and $P_{EIO}$ can be obtained as

$$P_{ISO}(\tau) = Pr\{S < S_d\} = 1 - PS(\tau) \text{ and } P_{EIO}(\tau) = Pr\{I > I_d\} = PI(\tau), \quad (6.58)$$

respectively. The expressions for $P_{ISO}$ and $P_{EIO}$ given in Eq. (6.48) and Eq. (6.49), respectively, can be used for general analysis of R-SLS detection algorithm.

## 6.5 RESULTS AND DISCUSSION

In this section, we carry out the experiments using theoretical analysis given in Section 6.4. The performance is illustrated using the plots of $P_{EIO}$ vs $P_{ISO}$ under Nakagami fading channels. The Monte Carlo simulations are also carried out in order to discuss the effect of different parameters on the performance of the proposed detection schemes. For Monte Carlo simulations, the results are averaged over $10^5$ realizations. Here, the plots shown in Figs. 6.5 to 6.13 are obtained using the mathematical expressions derived in Section 6.4 and those in Figs. 6.16, 6.17, and 6.18 are obtained using Monte Carlo simulations. Plots in Fig. 6.14 and Fig. 6.15 are obtained using both the approaches.

In Fig. 6.5, we display $P_{EIO}$ Vs $P_{ISO}$ for the proposed CC-SLC considering different number of diversity branches $P$. The plots are shown for $P$ ranging from 1 to 8. Note that $P = 1$ in CC-SLC corresponds to no diversity case, i.e., CCD. The value of $P_{ISO}$ for different values of $P$ with fixed value of $P_{EIO}$ of 0.2224 is tabulated in TABLE 6.2. We see that for a $P_{EIO} = 0.2224$, the $P_{ISO} = 0.0545$ for CCD and it is 0.01008 for CC-SLC with $P = 8$ indicating that $P_{ISO}$ is much smaller for given $P_{EIO}$ when we use diversity. This is illustrated in Fig. 6.5 using the dotted lines.

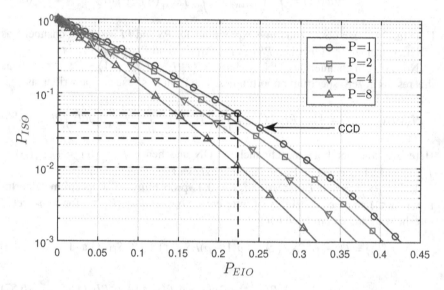

**Figure 6.5** $P_{EIO}$ Vs $P_{ISO}$ for CC-SLC considering $m_1 = 2, p = 0.1, L = 16, S_d = 1, I_d = 0, N = 10, \bar{\gamma} = -5$ $dB$ for different number of diversity branches $P$.

**Table 6.2**
**Effect of Increasing Number of Diversity Branches $P$ on $P_{ISO}$ for CC-SLC**

| Values of $P_{ISO}$ for different values $P$ with fixed $P_{EIO} = 0.2224$ | | | | |
|---|---|---|---|---|
| $P$ | 1 | 2 | 4 | 8 |
| $P_{ISO}$ | 0.0545 | 0.3846 | 0.2826 | 0.0108 |

In Fig. 6.6, we display $P_{EIO}$ VS $P_{ISO}$ for the proposed R-SLC considering different number of diversity branches $P$. The plots are shown for $P$ ranging from 1 to 8. Note that $P = 1$ in R-SLC corresponds to no diversity case, i.e., RCD. We have tabulated the values of $P_{ISO}$ for different values of $P$ with fixed $P_{EIO} = 0.04$ in TABLE 6.3. We see that for a $P_{EIO} = 0.040$, the $P_{ISO} = 0.5437$ for RCD and it is 0.2502 for R-SLC with $P = 8$ indicating that $P_{ISO}$ is much smaller for given $P_{EIO}$ when we use diversity. This is illustrated in Fig. 6.6 using the dotted lines.

**Table 6.3**
**Effect of Increasing Number of Diversity Branches $P$ on $P_{ISO}$ for R-SLC**

| Values of $P_{ISO}$ for different values $P$ with fixed $P_{EIO} = 0.04$ | | | | |
|---|---|---|---|---|
| $P$ | 1 | 2 | 4 | 8 |
| $P_{ISO}$ | 0.5437 | 0.4953 | 0.3879 | 0.2502 |

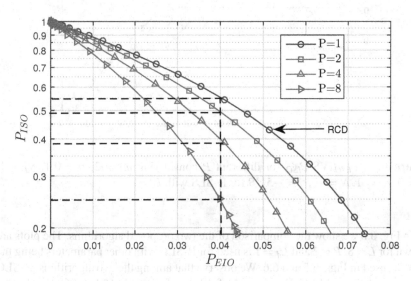

**Figure 6.6** $P_{EIO}$ VS $P_{ISO}$ for different $P$ considering $m_1 = 2$, $p = 0.1$, $S_d = 1$, $I_d = 0$, $L_d = 1$, $N = 10$, $\bar{\gamma} = -5\,dB$ for R-SLC with $L = 16$.

**Table 6.4**
**Effect of Increasing Number of Diversity Branches $P$ on $P_{ISO}$ for R-SLS**

| Values of $P_{ISO}$ for different values $P$ with fixed $P_{EIO} = 0.057$ | | | | |
|---|---|---|---|---|
| $P$ | 1 | 2 | 4 | 8 |
| $P_{ISO}$ | 0.4120 | 0.3857 | 0.3493 | 0.2751 |

Fig. 6.7 demonstrates the effect of increasing $P$ on the performance of R-SLS using the plots obtained by theoretical analysis. Once again $P$ is varied from 1 to 8

and the values of $P_{ISO}$ are tabulated in TABLE 6.4 for $P_{EIO} = 0.057$. We see that for a value of $P_{EIO} = 0.057$, the values of $P_{ISO}$ for RCD and R-SLC with $P = 8$ are 0.4120 and 0.2751, respectively, indicating the improvement in the performance when diversity is used. The dotted lines in Fig. 6.7 indicate this.

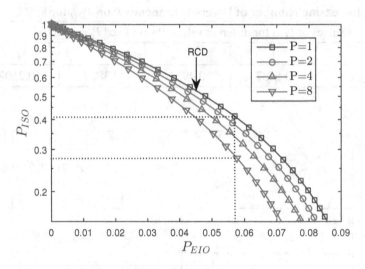

**Figure 6.7** $P_{EIO}$ VS $P_{ISO}$ for different $P$ considering $m_1 = 2$, $p = 0.1$, $S_d = 1$, $I_d = 0$, $L_d = 1$, $N = 10$, $\bar{\gamma} = -5$ $dB$ for R-SLS with $L = 2$.

In Fig. 6.8 we show the comparison of the two proposed algorithms. The plots are shown for $L = 8$, $P = 2$ and $L_d = 1$ is used for R-SLC with other parameters being the same as used in Figs. 6.5 and 6.6. We observe that among the two algorithms, R-SLC outperforms the CC-SLC. For example, for $P_{EIO}$ of approximately 0.062 we observe $P_{ISO} \approx 0.448$ for CC-SLC and 0.260 for R-SLC. This shows that $P_{ISO}$ for $P = 8$ is 41.96 % smaller than $P_{ISO}$ with $P = 1$. This is shown with dotted lines. This happens because in CC-SLC, we take the decision on all $L$ channels whereas in R-SLC we take the decision on only $L_d < L$ channels. Taking the decision on more number of channels results in increased interference. Since we are arranging the channels in the ascending order in R-SLC, the probability of finding free channels in first $L_d$ channels is high.

In Fig. 6.9 we show the comparison of the two proposed algorithms. These plots are obtained using the theoretical analysis in Section 6.4.2. The plots are shown for $L = 8$, $P = 2$ and $L_d = 1$ is used for R-SLC with other parameters being the same as used in Fig. 6.6. We observe that among the two algorithms, R-SLC performs better than R-SLS. For example, for $P_{EIO}$ of approximately 0.063 we observe $P_{ISO} \approx 0.260$ for R-SLC and 0.282 for R-SLS. This shows that $P_{ISO}$ for R-SLC is 7.8 % smaller than that for R-SLS algorithm.

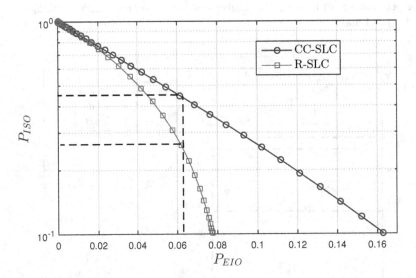

**Figure 6.8**  $P_{EIO}$ Vs $P_{ISO}$ showing the comparison of CC-SLC with R-SLC considering $m_1 = 2$, $p = 0.1$, $L = 8$, $S_d = 1$, $I_d = 0$, $L_d = 1$, $N = 10$, $\bar{\gamma} = -5\ dB$.

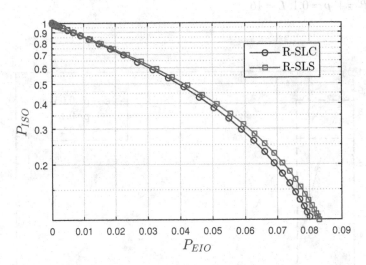

**Figure 6.9**  $P_{EIO}$ VS $P_{ISO}$ showing the comparison of R-SLC with R-SLS considering $m_1 = 2$, $p = 0.1$, $L = 2$, $P = 4$, $S_d = 1$, $I_d = 0$, $L_d = 1$, $N = 10$, $\bar{\gamma} = -5\ dB$.

We next show the effect of different parameters on the performance of the R-SLC detection algorithm. In Figs. 6.10, 6.11, and 6.12 we display the effect of average SNR ($\bar{\gamma}$), occupancy probability $p$ and the number of channels in the partial band $L$

on the performance of R-SLC. This analysis is done using $m_1 = 2$, $N = 10$, $S_d = 1$, $I_d = 0$, $L_d = 1$ and $P = 4$.

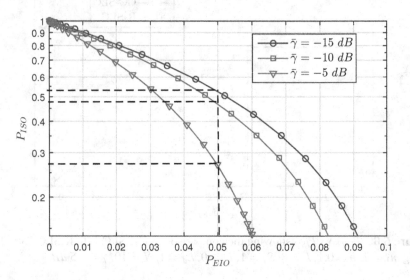

**Figure 6.10** $P_{EIO}$ VS $P_{ISO}$ for varying $\bar{\gamma}$ using $m_1 = 2$, $N = 10$, $S_d = 1$, $I_d = 0$, $L_d = 1$, $P = 4$, $p = 0.1$, $L = 16$.

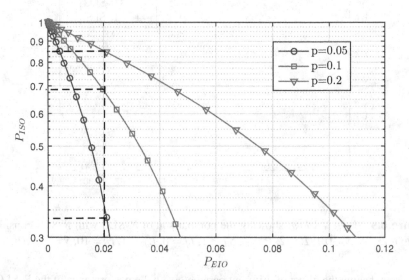

**Figure 6.11** $P_{EIO}$ VS $P_{ISO}$ for varying $p$ using $m_1 = 2$, $N = 10$, $S_d = 1$, $I_d = 0$, $L_d = 1$, $P = 4$, $\bar{\gamma} = -5\ dB$, $L = 16$.

# Diversity for Wideband Spectrum Sensing under Fading

### Table 6.5
### Effect of Increasing Average SNR $\bar{\gamma}$ on $P_{ISO}$ for R-SLC

| Values of $P_{ISO}$ for different values $\gamma$ with fixed $P_{EIO} = 0.05$ | | | |
|---|---|---|---|
| $\bar{\gamma}$ in dB | −15 | −10 | −5 |
| $P_{ISO}$ | 0.5321 | 0.4836 | 0.2690 |

### Table 6.6
### Effect of Increasing Occupancy Probability $p$ on $P_{ISO}$ for R-SLC

| Values of $P_{ISO}$ for different values $p$ with fixed $P_{EIO} = 0.02$ | | | |
|---|---|---|---|
| $p$ | 0.05 | 0.1 | 0.2 |
| $P_{ISO}$ | 0.3352 | 0.6881 | 0.8460 |

### Table 6.7
### Effect of Increasing Number of Channels in the Partial Band $L$ on $P_{ISO}$ for R-SLC

| Values of $P_{ISO}$ for different values $L$ with fixed $P_{EIO} = 0.04$ | | | | | | | |
|---|---|---|---|---|---|---|---|
| $L$ | 2 | 4 | 8 | 16 | 24 | 32 | 40 |
| $P_{ISO}$ | 0.5000 | 0.4603 | 0.4319 | 0.4069 | 0.3914 | 0.3879 | 0.3857 |

Fig 6.10 shows the effect of increasing $\bar{\gamma}$ for $p = 0.1$ and $L = 16$ where $\bar{\gamma}$ is increased from $-15$ dB to $-5$ dB. The values of $P_{ISO}$ obtained for $P_{EIO} = 0.05$ is tabulated in TABLE 6.5. It is observed that with the increase in $\bar{\gamma}$, the performance improves. It can be observed from the TABLE 6.5 that for $P_{EIO} = 0.05$ we get $P_{ISO} = 0.5321$ with $\bar{\gamma} = -15$ dB and $P_{ISO} = 0.2690$ with $\bar{\gamma} = -5$ dB. This indicates that $P_{ISO}$ with $\bar{\gamma} = -5$ dB is 49.45 % smaller than that with $\bar{\gamma} = -15$ dB. In Fig. 6.11, the effect of increase in $p$ of PUs is shown using $\bar{\gamma} = -5$ dB and $L = 16$. As $p$ increases the number of occupied channels in the partial band increases resulting in higher interference and lower spectrum opportunity and hence the performance degrades which is clearly seen in the plots. Also, the values of $P_{ISO}$ obtained for different values of $p$ for fixed value of $P_{EIO} = 0.02$ is tabulated in TABLE 6.6. We can clearly see that values of $P_{ISO}$ for $p = 0.05$ and $p = 0.2$ are 0.3352 and 0.8460 indicating that $P_{ISO}$ with $p = 0.05$ is 60.38 % smaller than that with $p = 0.2$. The effect of increasing $L$ is illustrated in Fig. 6.12. Here, the simulation is done by varying $L$ from 2 to 40 keeping $\bar{\gamma} = -5$ dB and $p = 0.1$. The values of $P_{ISO}$ for different values of $L$ with

$P_{EIO} = 0.04$ is tabulated in TABLE 6.7. We see that increase in $L$ results in improved performance. This is because increasing $L$ results in more number of free channels in the partial band. It can also be observed that when $L$ exceeds a certain value, the improvement is not significant thus validating the use of PBNS. Since the SUs do not require all the channels for use, it is not required to sense the entire wideband. For example, in Fig. 6.12, there is not much improvement in the performance for $L > 16$ and hence there is no advantage in sensing more than 16 channels.

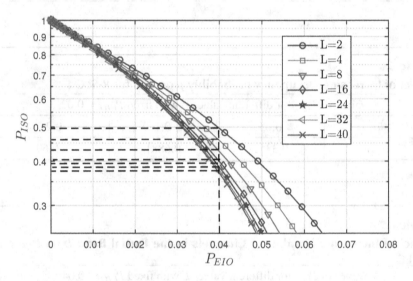

**Figure 6.12** $P_{EIO}$ VS $P_{ISO}$ for varying $L$ using $m_1 = 2$, $N = 10$, $S_d = 1$, $I_d = 0$, $L_d = 1$, $P = 4$, $\bar{\gamma} = -5\ dB$, $p = 0.1$.

**Table 6.8**
**Occupancy Probabilities for $L = 8$ Subbands**

| Occupancy Probability | $p_1$ | $p_2$ | $p_3$ | $p_4$ | $p_5$ | $p_6$ | $p_7$ | $p_8$ |
|---|---|---|---|---|---|---|---|---|
| All Low | 0.09 | 0.07 | 0.11 | 0.12 | 0.08 | 0.12 | 0.13 | 0.1 |
| Few High | 0.1 | 0.2 | 0.09 | 0.3 | 0.4 | 0.12 | 0.25 | 0.08 |
| All High | 0.2 | 0.25 | 0.3 | 0.15 | 0.4 | 0.35 | 0.2 | 0.45 |

In Fig. 6.13, we show the $P_{EIO}$ VS. $P_{ISO}$ plots considering different $p$ for primary channels. These plots are obtained for three scenarios namely: 1. all the channels have low occupancy probability, 2. only few channels have high occupancy probability and 3. all channels have high occupancy probability. Different $p$'s considered for $L = 8$ are given in TABLE 6.8. We observe that the performance is better when all the channels have lower p and the performance degrades when all of them have high p. For example, for $P_{EIO}$ of approximately 0.020 we observe $P_{ISO}$ as approximately 0.616, 0.804, and 0.881 for three cases of all low, few high and all high, respectively. These values are tabulated in TABLE 6.9 for reference. From this one can deduce that those groups of channels where occupancy probabilities are not significantly high have to be selected to yield better sensing performance.

**Table 6.9**
**Effect of Different Occupancy Probabilities on $P_{ISO}$ for R-SLC**

| Values of $P_{ISO}$ for different occupancy probabilities with fixed $P_{EIO} = 0.02$ | | | |
|---|---|---|---|
| p | All Law | Few Law | High |
| $P_{ISO}$ | 0.616 | 0.804 | 0.881 |

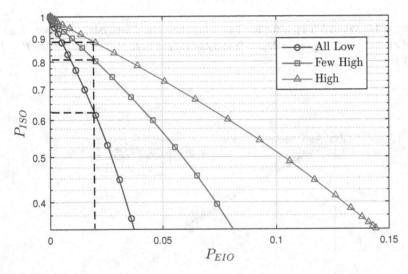

**Figure 6.13** $P_{EIO}$ VS $P_{ISO}$ plots considering different p for channels in the partial band for $m_1 = 2, L = P = 8, S_d = 1, I_d = 0, L_d = 1, N = 10, \bar{\gamma} = -5\ dB$.

In Fig. 6.14 and Fig. 6.15, we show the plots obtained using theoretical expressions derived in Section 6.4 and using the Monte Carlo simulations. We see that the plots almost overlap, validating the theoretical analysis carried out for both R-SLC and R-SLS. We can see that there is a slight deviation of the simulated plot from the theoretical plot. This is because of the approximation we have used in deriving the pdf of decision statistics under SLC and SLS diversity.

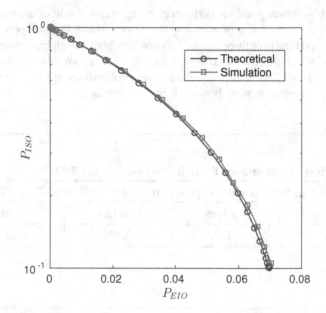

**Figure 6.14** $P_{EIO}$ VS $P_{ISO}$ for R-SLC using theoretical analysis and Monte Carlo simulation considering $m_1 = 2$, $p = 0.1$, $S_d = 1$, $I_d = 0$, $L_d = 1$, $N = 10$, $L = P = 4$, and $\bar{\gamma} = -5\ dB$.

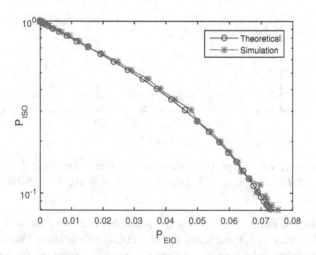

**Figure 6.15** $P_{EIO}$ VS $P_{ISO}$ for R-SLS using theoretical analysis and Monte Carlo simulation considering $m_1 = 2$, $p = 0.1$, $S_d = 1$, $I_d = 0$, $L_d = 1$, $N = 10$, $P = L = 2$, and $\bar{\gamma} = 0\ dB$.

Finally, in Figs. 6.16, 6.17, and 6.18 we show the effect of $L_d$, $S_d$, and $I_d$ on the performance of R-SLC using Monte Carlo simulations. Fig. 6.16 shows the effect of increasing $L_d$ by considering $L = 8$ and keeping the other parameters fixed. The values of $P_{ISO}$ for fixed value of $P_{EIO} = 0.03$ is tabulated in TABLE 6.10 for different values of $L_d$. We see that increasing $L_d$ degrades the performance. For example, we get $P_{ISO} = 0.3199$ with $L_d = 3$ and $P_{ISO} = 0.0975$ with $L_d = 1$ indicating that $P_{ISO}$ with $L_d = 3$ is 228.1% higher than that with $L_d = 1$. This is because although increasing $L_d$ enhances the chance of getting free channels it also increases interference to the PUs. With $I_d = 0$, increasing $L_d$, degrades the performance. From this result, we can say that it is advantageous to choose $L_d = S_d$.

**Table 6.10**
**Effect of Increasing $L_d$ on $P_{ISO}$ for R-SLC**

| | Values of $P_{ISO}$ for different value $L_d$ with fixed $P_{EIO} = 0.03$ | | |
|---|---|---|---|
| $L_d$ | 1 | 2 | 3 |
| $P_{ISO}$ | 0.0975 | 0.3021 | 0.3199 |

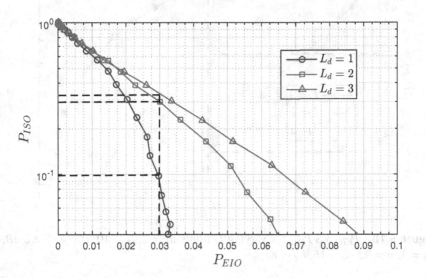

**Figure 6.16** $P_{EIO}$ VS $P_{ISO}$ for varying $L_d$ using $m_1 = 2$, $N = 10$, $P = 4$, $\bar{\gamma} = 0\,dB$, $S_d = 1$, $I_d = 0$, $p = 0.1$, $L = 8$.

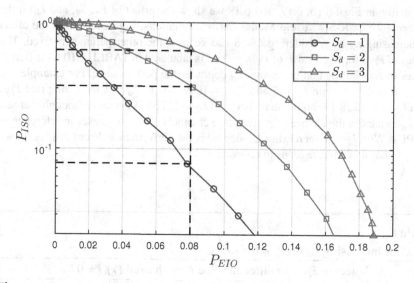

**Figure 6.17** $P_{EIO}$ VS $P_{ISO}$ for varying $S_d$ using $m_1 = 2, N = 10, P = 4, \bar{\gamma} = 0\ dB$, $L_d = 4, p = 0.1, L = 8, I_d = 0$.

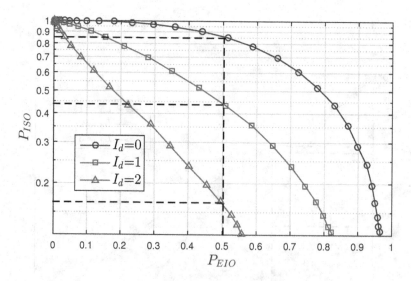

**Figure 6.18** $P_{EIO}$ VS $P_{ISO}$ for varying $I_d$ using $m_1 = 2, N = 10, P = 4, \bar{\gamma} = 0\ dB$, $S_d = 4, p = 0.5, L = 16, L_d = 8$.

Next, we illustrate the effect of an increase of $S_d$ in Fig. 6.17. For this simulation, we consider $S_d = 1, 2$ and $3$ using $m_1 = 2, N = 10, I_d = 0, \bar{\gamma} = 0\ dB, p = 0.1$,

Diversity for Wideband Spectrum Sensing under Fading

$P = 4$, $L_d = 4$ and $L = 8$. As seen from the plots, it is clear that as $S_d$ increases, the performance degrades. The values of $P_{ISO}$ considering different values of $S_d$ with fixed $P_{EIO} = 0.08$ is tabulated in TABLE 6.11. From the table, we can see that $P_{ISO}$ value is greatly affected by the increase in the value of $S_d$. This happens because, if we increase $S_d$, it requires $L_d$ to be increased since $S_d \leq L_d$. However, increasing $L_d$ results in making decisions on more channels leading to higher interference to the PUs causing degradation in the performance. Finally, in Fig. 6.18, we demonstrate the effect of varying $I_d$ on the performance. The values of $P_{ISO}$ obtained for fixed $P_{EIO} = 0.5$ is tabulated in TABLE 6.12. We can see that performance improves by increasing the values of $I_d$. Increasing $I_d$ (i.e., $I_d > 0$) implies SUs are allowed to interfere with the PUs. For example, with $I_d = 1$, interference to any one of the PU channels is allowed. Although we observe that performance improves as $I_d$ increases, it is not desirable to allow interference to the primary users. Note that, we have given plots showing the effect of different parameters on the performance of R-SLC, only. One can also show similar plots for R-SLS. All the conclusions drawn for R-SLC are also valid for R-SLS.

**Table 6.11**
**Effect of Different $S_d$ on $P_{ISO}$ for R-SLC**

| Values of $P_{ISO}$ for different values of $S_d$ with fixed $P_{EIO} = 0.08$ | | | |
|---|---|---|---|
| $S_d$ | 1 | 2 | 3 |
| $P_{ISO}$ | 0.07625 | 0.3088 | 0.6145 |

**Table 6.12**
**Effect of Different $I_d$ on $P_{ISO}$ for R-SLC**

| Values of $P_{ISO}$ for different occupancy probabilities with fixed $P_{EIO} = 0.5$ | | | |
|---|---|---|---|
| $I_d$ | 0 | 1 | 2 |
| $P_{ISO}$ | 0.8509 | 0.434 | 0.1665 |

## 6.6 CONCLUSION

In this chapter, new detection algorithms are discussed for wideband spectrum sensing that use diversity to improve the detection performance. A complete mathematical analysis is given to find the performance of wideband spectrum sensing with and without diversity by considering Nakagami fading. The analysis given is general and can be used with any fading model and any diversity technique. From the analysis, it is clear that the use of diversity improves detection performance. The effect of different parameters on the performance is demonstrated using plots. We observed that R-SLC technique outperforms both CC-SLC and R-SLS. The future extension of this work involves using other diversity schemes and more general fading models, including frequency selective fading for analyzing WSS's performance. Next, an optimization technique can be used for finding the optimal thresholds when different priorities and different occupancy probabilities are considered for PU channels. Furthermore, the approach in this chapter can be extended to cooperative wideband spectrum sensing where multiple secondary users cooperate in detecting the presence of the primary users.

# 7 Cooperative Wideband Spectrum Sensing

In the previous chapters, we discussed antenna diversity for wideband spectrum sensing to improve the detection performance. When we use antenna diversity, the spacing between antennas has to be sufficiently high to receive independent observations, which limits the number of antennas that can be used at the secondary users. Hence, makes it difficult to use higher number of diversity branches at the SUs. Though it is advantageous to have diversity at the SUs, one cannot increase the number of diversity branches because of the spacing requirement between the antennas. The cooperative spectrum sensing has been shown as an effective method that improves the detection performance by exploiting the spatial diversity [7]. In cooperative spectrum sensing number of cooperating secondary users collaborate to detect the presence or the absence of the primary user. The information collected by different CSUs are utilized to make decision on the occupancy of primary user channels. There are number of ways to implement CSS as discussed in Chapter 2. In this chapter, an algorithm using the centralized CSS for wideband spectrum sensing (WSS) is discussed. There are some works attempted in literature for cooperative wideband spectrum sensing (CWSS). An expectation maximization based joint detection and estimation (JDE) scheme for cooperative spectrum sensing in multiuser multiantenna CR network is proposed in [16], wherein multiple spatially separated SUs cooperate to detect the state of occupancy of a wideband frequency spectrum. In [155], the spectrum sensing problem is formulated as a class of optimization problems that maximize the aggregated opportunistic throughput of a CR system under the constraints of interference to the primary users. All these techniques use traditional sampling techniques for wideband sensing and hence require higher sampling rates. This chapter proposes two new algorithms based on cooperative spectrum sensing with hard and soft combining for wideband spectrum sensing. Since the idea of partial band Nyquist sampling is not explored in CSS, we make use of PBNS at all the CSUs to reduce the higher sampling rate requirements. The difficulty of having multiple antennas at the SU is overcome by using CSS with fewer diversity branches at each cooperating secondary users. For example, one may increase the number of CSUs but restrict the number of diversity branches to two at each CSU to enhance the detection performance. Here, in this work, a PBNS based detection algorithms for cooperative WSS are discussed using cooperation from a number of SUs equipped with two diversity branches only. In effect, the algorithms use both space and antenna diversities to improve performance. The hard combining for data fusion is used in which all the CSUs take decisions on the occupancy of the channels and send the final results to the fusion center (FC), which takes the final decision. Although soft combining may result in better detection performance, it requires higher control channel bandwidth compared to hard combining, which is a crucial drawback of soft combining considering the paucity of bandwidth. The algorithm based on soft combining is also discussed in this chapter. The cooperating spectrum sensing based hard combining is a well-researched area in spectrum sensing. However, it may be noted that all the available works using hard combining are restricted to narrowband sensing only, and the researchers have not yet explored the same for WSS as done here. Our approach uses PBNS as the wideband sampling scheme at each CSU, and the performance is measured in terms of the probability of insufficient spectrum opportunity ($P_{ISO}$) and the probability of excessive interference opportunity ($P_{EIO}$). A complete theoretical analysis is carried out considering the Nakagami fading channel, which is a generalized fading model and includes Rayleigh, Rice, and Hoyt fading models as its special cases. It can be used to model propagation in urban and sub-urban areas. The experiments are carried out using theoretical analysis and also verified using Monte Carlo (MC) simulations. The analysis given in the chapter shows that the proposed approaches outperform the ranked channel detection used under no cooperation. It is also demonstrated that by choosing an appropriate number of CSUs, the algorithm performs better than both R-SLC and R-SLS algorithms discussed in Chapter 6.

## 7.1 SYSTEM MODEL AND PERFORMANCE METRICS

The wideband signal is modeled as the collection of $U$ subbands, i.e., narrowband channels, each of bandwidth $B_0$ making the total bandwidth as $B = UB_0$. Similar channel model is used in [14, 36, 185]. Here, the primary transmission within each subband is subjected to flat fading. Let $H_m$ denotes the primary occupancy of the $m^{th}$ subband, with $H_m = 0$ and $H_m = 1$ corresponding to PU being OFF (free) and ON, respectively. In our discussions, we assume that the occupancy status of the PU remains unchanged during the observation interval. This assumption is valid as long as the observation is not too long to make it comparable to the primary ON/OFF time. It is also assumed that all the PUs have equal priorities. In this chapter, the discussion is restricted to one sensing window with a fixed duration of $T_w = UNT_N$, where $N$ represents the number of samples per subband, $T_N$ is the sampling duration, and $F_N = 1/T_N = 2B$ is the Nyquist sampling frequency. We want to mention here that the terms subband and channel are used interchangeably. It is assumed that the occupancy status of subbands and their transmissions are independent of each other. Assuming the occupancy probability of subbands as $p$, the state of each subband is modeled by a Bernoulli random variable and the probability mass function (PMF) of the same given as

$$f(H_m, p) = \begin{cases} p, & \text{if } H_m = 1 \\ 1-p, & \text{if } H_m = 0. \end{cases} \quad (7.1)$$

With the number of ON channels as $K$, PMF of primary occupancy is given by Binomial distribution as

$$f(K, U, p) = \binom{U}{K} p^K (1-p)^{U-K}. \quad (7.2)$$

Note that in order to detect the free primary channels the detector has to decide on their presence or absence. The received signal $y_m(t)$ at the SU due to $m^{th}$ primary subband by considering flat fading can now be modeled as

$$y_m(t) = h_m \cdot s_m(t) + n_m(t), \qquad (7.3)$$

where, $h_m$ is the channel coefficient for $m^{th}$ subband which is assumed to be Nakagami distributed, $s_m(t)$ and $n_m(t)$ represent the transmitted signal and the additive white Gaussian noise in $m^{th}$ primary subband, respectively.

Two new performance metrics, i.e., the probability of insufficient spectrum opportunity ($P_{ISO}$) and the probability of excessive interference opportunity ($P_{EIO}$), are proposed in [185], which are better suited for WSS and take into account the required number of spectrum opportunities ($S_d$) and the allowed interference to the PU channels ($I_d$). We make use of these to evaluate the performance of our proposed algorithm. Let $S$ and $I$ represent the number of SO, i.e., the number of successfully identified OFF channels, and the number of missed ON channels, respectively. Then, $P_{ISO}$ and $P_{EIO}$ are defined as

$$P_{ISO}(S_d) = Pr\{S < S_d\} \text{ and } P_{EIO}(I_d) = Pr\{I > I_d\}, \qquad (7.4)$$

where, $Pr$ represents probability.

## 7.2 PROPOSED CWSS ALGORITHMS

Consider the $M$ number of CSUs receiving the wideband signal consisting of $U$ channels. Since the SU may not be interested in finding all the spectrum opportunities, instead, the interest lies in finding sufficient number of spectrum opportunities; the idea of PBNS is used, which samples only a fraction of the entire wideband. The PBNS is characterized by the number of channels in the partial band ($L$) and filters out $L$ channels from the wideband having $U$ channels. The signal after filtering is sampled at the Nyquist rate of $2LB_0$ on which the fast Fourier transform (FFT) is performed. Taking $N$ samples per channel, the number of samples in the sensing window becomes $NL$. Considering $V[k]$ as the frequency samples in PBNS, we can write $V[k] = Y[k]$, for $0 \le k \le NL - 1$, where $Y[k]$ denotes the normalized discrete Fourier transform (DFT) of the received wideband signal having $M$ channels. The energy, i.e., decision statistic, within $m^{th}$ narrowband channel is then calculated as

$$T_m = \sum_{k \in I_m} |V[k]|^2, \qquad (7.5)$$

where, $1 \le m \le L$ and $I_m$ is the set of frequency indices that fall into channel m.

### 7.2.1 PROPOSED ALGORITHM BASED ON HARD COMBINING

In CSS, the number of SUs cooperate in order to detect spectrum holes. In the proposed algorithm, CSS with hard combining is used where all the CSUs take their own decision on the occupancy of the channels and send their final result to the FC, which takes the final decision. Fig. 7.1 shows the block schematic of the proposed detection algorithm, and the steps involved in implementing it are given in Algorithm 4. We consider that each CSU is equipped with two diversity branches and employs square law combining (SLC) diversity. With this, the decision statistic at the $i^{th}$ CSU for the $m^{th}$ PU channel can be obtained as

$$T^i_{m,slc} = T^i_{m,1} + T^i_{m,2}, \qquad (7.6)$$

where, $T^i_{m,1}$ and $T^i_{m,2}$ represent the decision statistic obtained at the $i^{th}$ CSU for the $m^{th}$ PU channel at diversity branches 1 and 2, respectively, which can be obtained using Eq. (7.5). Here, $T^i_{m,slc}$ represents the decision statistic obtained for the $m^{th}$ channel at the $i^{th}$ CSU using SLC.

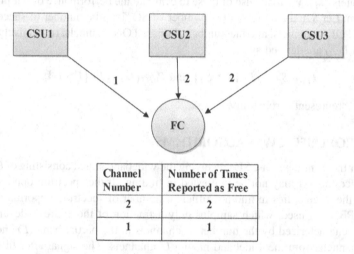

**Figure 7.1** Block diagram of proposed detection algorithm with $L = 2$, $M = 3$, $S_d = L_d = X_d = 1$, $I_d = 0$, $F_d = 1$.

Fig. 7.1, the case of $L = 2$ and $M = 3$ is discussed. The two channels in the partial band are numbered as channels 1 and 2, respectively. Let $L_d$ and $X_d$ represent the number of channels on which the FC and CSUs take the decision. Let $F_d$ represents the total number of times a channel must be reported as OFF by the CSUs in order for it to be declared as free. Considering $S_d = L_d = X_d = 1$, $I_d = 0$, the proposed detection scheme works as follows. After receiving the wideband signal, all the three CSUs perform PBNS at both the diversity branches. Each CSU then computes the decision statistic using SLC diversity (see Eq. (7.6)) for both the channels and performs ranked channel detection (RCD), i.e., arrange $T^i_{1,slc}$ and $T^i_{2,slc}$ in ascending order and then take a decision on the first channel only by comparing the decision statistic with

# Cooperative Wideband Spectrum Sensing

a chosen threshold $\tau$. If this channel is declared as free, the CSUs send the channel number to the FC, which computes how many times a particular channel is reported as free. The FC then arranges them in descending order with regard to the number of times a particular channel is reported as free. In Fig. 7.1, channel 1 is reported as free once, whereas the channel 2 is reported as free twice. With $S_d = L_d = 1$, the FC takes a decision on that channel, which appears on the top of the order. As shown in Fig. 7.1, channel 2 is reported as free twice which is greater than $F_d = 1$ and hence the FC declares channel 2 as free. If both the channels are reported free for an equal number of times and are reported more than $F_d$ times, the FC chooses any one channel from them with equal probability. For example, with $M = 4$, if both channels are reported free twice, the FC declares any one of them as free with a probability of 0.5.

**Algorithm 4**

1: At each CSU: Compute decision statistic using SLC diversity (Eq. 7.6) for each of the $L$ channels in the partial band and arrange them in ascending order. Take decision on first $X_d$ channels.
2: Each CSU reports the free channel numbers to the fusion center (FC).
3: FC counts the number of times a particular channel is reported free and then arranges them in descending order with regard to it.
4: At FC: Take decision on first $L_d$ channels, where $S_d \leq L_d \leq X_d \leq L$, and report the channels as free if they are reported free $> F_d$ times. If channels are reported free for equal number of times and are reported free $> F_d$ times, the FC declares them as free with equal probability.

## 7.2.2 PROPOSED ALGORITHM BASED ON SOFT COMBINING

In CSS, the number of SUs cooperate in order to detect spectrum holes. The pictorial representation of the proposed algorithm using soft combining is shown in Fig. 7.2. We next discuss the algorithm with soft combining. Following are the steps involved in the algorithm. First, the CSUs compute decision statistics for $L$ channels in the partial band and send them to the FC which makes the final decision. Let us consider that all the CSUs have $P$ number of diversity branches and make use of square law combining (SLC) diversity. The combined decision statistic is then obtained in SLC diversity by adding the measured decision statistic from different diversity branches. Hence, the decision statistic after employing SLC at the $i^{\text{th}}$ CSU for the $m^{\text{th}}$ channel is obtained as

$$T_{m,slc}^i = \sum_{j=1}^{P} T_{m,j}^i, \qquad (7.7)$$

where, $T_{m,j}^i$ is the decision statistic computed at the $i^{\text{th}}$ CSU for the $m^{\text{th}}$ channel at $j^{\text{th}}$ diversity branch that can be computed using Eq. (7.5). Note that, if we choose $P = 2$ in Eq. (7.7), we get the Eq. (7.6). All the CSUs report the computed decision statistic to the FC which then adds all the received decision static for each channel from

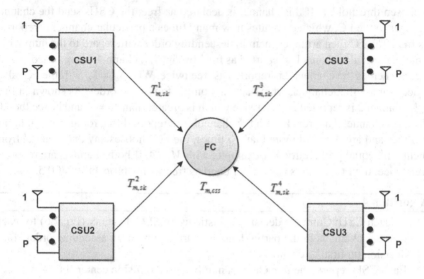

**Figure 7.2** Block diagram for the proposed detection scheme with four CSUs ($M = 4$).

all CSUs to obtain new decision statistic. With $M$ number of CSUs, new decision statistic at the FC for $m^{th}$ channel is obtained as

$$T_{m,css} = \sum_{i=1}^{M} T_m^i. \tag{7.8}$$

The FC then arranges these decision statistics in ascending order and takes decision on first $L_d$ channels by comparing them with the threshold $\tau$. The steps involved in the proposed algorithm are listed in Algorithm 5.

---

**Algorithm 5**

---

1: At each CSU: Compute decision statistic using SLC diversity (Eq. 7.7) for each of the $L$ channels in the partial band.
2: Each CSU reports the decision statistic for each of the $L$ channels to the fusion center (FC).
3: The FC adds received decision statistics for each channel from $M$ CSUs to obtain new decision statistics (Eq. (7.8)) for all channels.
4: At FC: Arrange new decision statistics in ascending order and take decision on first $L_d$ number of channels where $S_d \leq L_d \leq L$.

---

## 7.3  APPROXIMATION TO PDF OF DECISION STATISTIC

The pdf of decision statistic, i.e., received energy $T_m$, for the $m^{\text{th}}$ channel under Nakagami fading is given by [105],

$$f_{T_m}(t) = \begin{cases} \frac{1(t)t^{G-1}e^{-\frac{t}{2}}}{\Gamma(G)2^G D} {}_1F_1(m_1; G; tE), & H_m = 1 \\ \chi_N^2, & H_m = 0 \end{cases} \quad (7.9)$$

where, $D = (1 + \bar{\gamma}/m_1)^{m_1}$, $E = 0.5 - 0.5m_1/(m_1 + \bar{\gamma})$, $G = N/2$. Here, $\bar{\gamma}$ is the average signal to noise ratio at the input of SU, $1(t)$ is the unit step function, $m_1$ is the Nakagami parameter, ${}_1F_1(\cdot;\cdot;\cdot)$ represents the confluent hypergeometric function [79] and $\chi_N^2$ is a central chi-square distribution with degree of freedom $N$.

One can use the pdf given Eq. (7.9) in order to derive performance metrics, i.e., $P_{ISO}$ and $P_{EIO}$, but this leads to infinite series for the final expressions and makes it very difficult to draw insights into the effects of different parameters on the performance. One can overcome this issue by approximating the pdf given in Eq. (7.9) under $H_m = 1$ as done Chapter 6 (Section 6.3), which lacks in better approximating the pdf. The problem with the approximation given in Chapter 6 is that only the first two terms of the Maclaurin's series expansion of the original pdf given in Eq. (7.9) under $H_m = 1$ is used. In this chapter, we first derive a better approximation to the pdf and then use it in subsequent analysis. As done in Chapter 6, using Maclaurin's series expansion of a function $f(t)$, one can write

$$f(t) = \sum_{i=0}^{\infty} a_i t^{q+i}, \text{ as } t \to 0^+. \quad (7.10)$$

The authors in [26] propose the approximation as

$$f(t) = at^q \left( e^{-\theta_1 t} + e^{-\theta_2 t} \right), \quad (7.11)$$

where, $a = a_0/2$, $\theta_1 = (b_1 + \sqrt{2b_2 - b_1^2})/2$, $\theta_2 = (b_1 - \sqrt{2b_2 - b_1^2})/2$, $b_1 = -2a_1/a_0$ and $b_2 = 4a_2/a_0$.

The Maclaurin's series expansion of Eq. (7.9) under $H_m = 1$ can be expressed using first three terms as

$$f_{T_m}(t) = \frac{1}{2^G D \Gamma(G)} t^{G-1} + \frac{(2Em_1 - G)}{2^{G+1} D \Gamma(G+1)} t^G$$
$$+ \frac{(1+G)(G - 4Em_1) + 4E^2 m_1 (m_1 + 1)}{2^{G+3} D \Gamma(G+2)} t^{G+1} + O(t^{G+2}). \quad (7.12)$$

On comparing Eq. (7.12) with Eq. (7.10), we get

$$q = G - 1, \quad a_0 = \frac{2^{-G}}{D \Gamma(G)}, \quad a_1 = \frac{(2Em_1 - G)}{2^{G+1} D \Gamma(G+1)},$$

$$a_2 = \frac{(1+G)(G-4Em_1)+4E^2m_1(m_1+1)}{2^{G+3}D\Gamma(G+2)}.$$

Once we get $a_0$, $a_1$ and $a_2$, we can compute $\theta_1$ and $\theta_2$. With this, the approximated pdf of $T_m$ is given by

$$f_{T_m}(t) \approx \begin{cases} at^q\left(e^{-\theta_1 t}+e^{-\theta_2 t}\right), & H_m = 1 \\ \chi_N^2. & H_m = 0 \end{cases} \quad (7.13)$$

Note that, though this approximation may not result in a proper pdf, i.e., the area under $f_{T_m}(t)$ is not necessarily 1, it gives better approximation than that used in Chapter 6. In Fig. 7.3, we show the plots for pdf showing the comparison of the proposed approximation given in Eq. (7.13) under $H_m = 1$ and that used in Chapter 6. We also compare the approximation with the true pdf given in Eq. (7.9). We can clearly see that the proposed approximation is better than the one used in Chapter 6. As shown in Fig. 7.3a for $\bar{\gamma} = -5\ dB$, the plot obtained using the approximation in Chapter 6 nearly overlaps with the true pdf, but the proposed approximation completely overlaps with the true pdf. Also, as $\bar{\gamma}$ increases, the approximation of Chapter 6 starts deviating from the true pdf, but the proposed approximation still fallows it. In Fig. 7.3b, we show the plots at $\bar{\gamma} = 0\ dB$. We also see that the proposed approximation almost overlaps, but the approximation used in Chapter 6 deviates significantly from the true pdf. The area under the pdf curve in Fig. 7.3b when the proposed approximation is used is 0.9919, whereas, that with the approximation in Chapter 6 is 0.9090, which shows a better approximation to the pdf in Eq. (7.9). We make use of this approximation for carrying out our theoretical analysis.

In the proposed algorithm, it is assumed that all the CSUs employ two diversity branches and use SLC diversity, in which the decision statistic received at the two diversity branches are added to derive a new decision statistic. Since the energy received at the two diversity branches are statistically independent, the pdf of the sum can be obtained by convolving their marginal pdfs. Assuming that both the diversity branches have the same average SNRs ($\bar{\gamma}$) [60], the pdf under $H_m = 1$ when SLC is used at the CSUs can be obtained by convolving the pdfs in Eq. (7.13) under $H_m = 1$ with itself and the same is given by Eq. (7.14). In Eq. (7.14), $I_n(x)$ represents the modified Bessel function of the first kind. Note that, under $H_m = 0$, adding two independent and identically distributed (iid) central chi-square random variables, each with $N$ degrees of freedom (DOF), results in another chi-square random variable with $2N$ DOF.

$$f_{T_m}(t) \approx \begin{cases} \dfrac{a^2\Gamma(1+q)^2 t^{2q+1}\left(e^{-\theta_1 t}+e^{-\theta_2 t}\right)}{\Gamma(2q+2)} \\ +a^2\sqrt{\pi}\Gamma(1+q)t^{q+\frac{1}{2}}e^{-\left(\frac{\theta_1+\theta_2}{2}\right)t}\left[\dfrac{I_{q+\frac{1}{2}}\left(\frac{\theta_1-\theta_2}{2}t\right)}{(\theta_1-\theta_2)^{q+\frac{1}{2}}}+\dfrac{I_{q+\frac{1}{2}}\left(\frac{\theta_2-\theta_1}{2}t\right)}{(\theta_2-\theta_1)^{q+\frac{1}{2}}}\right], & H_m=1 \\ \chi_{2N}^2. & H_m=0 \end{cases}$$
$$(7.14)$$

Cooperative Wideband Spectrum Sensing

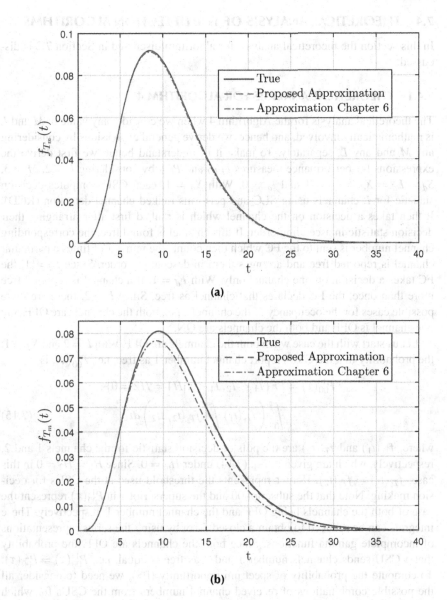

**Figure 7.3** Plots showing the true and the approximated pdfs under $H_m = 1$ considering $N = 10$, $m_1 = 2$ for (a) $\bar{\gamma} = -5 \ dB$ and (b) $\bar{\gamma} = 0 \ dB$.

## 7.4 THEORETICAL ANALYSIS OF THE DETECTION ALGORITHMS

In this section the theoretical analysis for algorithm discussed in Section 7.2 is discussed.

### 7.4.1 THEORETICAL ANALYSIS FOR ALGORITHM 4

The theoretical analysis for the Algorithm 4 when we consider any value of $M$ and $L$ is mathematically involved, and hence, we derive general expressions by considering any $M$ and any $L$, separately. To make it to understand better, we first derive the expressions for performance measures $P_{EIO}$ and $P_{ISO}$ by considering $L = 2$, $M = 3$, $S_d = L_d = X_d = F_d = 1$ and $I_d = 0$. With $X_d = 1$, each CSU computes decision statistic for $L$ channels using SLC and performs ranked channel detection (RCD). It then takes a decision on the channel which is ranked first after arranging their decision statistic in ascending order. If this channel is found free, the corresponding channel number is sent to the FC which then counts the number of times a particular channel is reported free and arranges them in descending order. Since $L_d = 1$, the FC takes a decision on one channel only. With $F_d = 1$, if a channel is reported free more than once, the FC declares that channel as free. Since $L = 2$, there are three possible cases for the occupancy of the channels, i.e., both the channels are OFF, any one channel is OFF and both the channels are ON.

Let us start with the case when both the channels are OFF. With $L = 2$ and $X_d = 1$, the probability that the CSUs report channel number 1 as free, i.e., $P_{00}^1(\tau)$, is

$$P_{00}^1(\tau) = Pr\{T_1 < T_2, T_1 < \tau | H1 = H2 = 0\},$$

$$= \int_0^\tau \left( f_{T_1}(t_1) \int_{t_1}^\infty f_{T_2}(t_2) dt_2 \right) dt_1, \qquad (7.15)$$

where, $f_{T_1}(t_1)$ and $f_{T_2}(t_2)$ are the pdfs of decision statistic for the channels 1 and 2, respectively, which are given in Eq. (7.14) under $H_m = 0$. Since $H_1 = H_2 = 0$ in this case, $f_{T_1}(t_1) = f_{T_2}(t_2)$. Here $\tau$ represents the threshold used at the CSUs for decision making. Note that the subscript 00 and the superscript 1 in $P_{00}^1(\tau)$ represent the case of both the channels being OFF and the channel number 1, respectively. These integrals can be reduced to obtain a closed form by using the series representations of incomplete gamma functions. Since both the channels are OFF, the probability that a CSU sends channels number 1 and 2 as free is equal, i.e., $P_{00}^1(\tau) = P_{00}^2(\tau)$. To compute the probability of spectrum opportunity (PS), we need to consider all the possible combinations of received channel numbers from the CSUs for which the FC declares any one channel as free. For example, one possible combination of received free channel numbers could be CSU1 and CSU2 sending channel number 1, and CSU3 sending channel number 2. In this case, the FC declares channel 1 as free. Note that with $F_d = 1$, only two CSUs reporting the same channel number as free is sufficient to declare that channel as free. Considering this, probability of spectrum opportunity in this case, i.e., $PS_{00}$, can be obtained as

$$PS_{00}(\tau) = 8\left[P_{00}^1(\tau)\right]^3 + 6\left[P_{00}^1(\tau)\right]^2 \left[1 - 2P_{00}^1(\tau)\right]. \qquad (7.16)$$

Since there is no chance of interference in this case, the probability of interference (PI) is zero, i.e., $PI_{00}(\tau) = 0$.

We now consider the case when only one channel is ON. Let us assume that channel 1 is OFF and channel 2 is ON. In this case, the probability that a CSU reports channel 1 as free, i.e., $P_{01}^1(\tau)$, can be obtained using Eq. (7.15). Since $H_1 = 0$ and $H_2 = 1$, $f_{T_1}(t_1)$ and $f_{T_2}(t_2)$ are to be used from Eq. (7.14) under $H_m = 0$ and $H_m = 1$, respectively. The probability that a CSU reports channel 2 as free, i.e., $P_{01}^2(\tau)$, can be obtained as

$$P_{01}^2(\tau) = Pr\{T_2 < T_1, T_2 < \tau | H1 = 0, H2 = 1\},$$
$$= \int_0^\tau \left( f_{T_2}(t_2) \int_{t_2}^\infty f_{T_1}(t_1) dt_1 \right) dt_2. \quad (7.17)$$

Since $H_1 = 0$ and $H_2 = 1$, $f_{T_1}(t)$ and $f_{T_2}(t)$ are to be used from Eq. (7.14) under $H_m = 0$ and $H_m = 1$, respectively.

In this case, we get the spectrum opportunity if the channel 1 is reported as free at least two times. For example, we get spectrum opportunity when CSUs 1 and 2 report channel number 1 as free and CSU3 reports channel 2 as free. Considering all such possible combinations, $PS_{01}(\tau)$ can be obtained as

$$PS_{01}(\tau) = \left[P_{01}^1(\tau)\right]^3 + 3\left[P_{01}^1(\tau)\right]^2 P_{01}^2(\tau) + 3\left[P_{01}^1(\tau)\right]^2 \left[1 - P_{01}^1(\tau) - P_{01}^2(\tau)\right]. \quad (7.18)$$

Similarly, it results in interference if the channel 2 is reported as free at least two times. In this case, the probability of interference, i.e., $PI_{01}$, can be obtained as

$$PI_{01}(\tau) = \left[P_{01}^2(\tau)\right]^3 + 3\left[P_{01}^2(\tau)\right]^2 P_{01}^1(\tau) + 3\left[P_{01}^2(\tau)\right]^2 \left[1 - P_{01}^2(\tau) - P_{01}^1(\tau)\right]. \quad (7.19)$$

Note that, if we consider channel 1 as ON and channel 2 as OFF, we get the same results, i.e., $PS_{01}(\tau) = PS_{10}(\tau)$ and $PI_{01}(\tau) = PI_{10}(\tau)$.

Finally, we consider the case when both the channels are ON. In this case there is no chance of spectrum opportunity and hence $PS_{11}(\tau) = 0$. The probability of reporting channel 1 as free, i.e., $P_{11}^1(\tau)$, can be obtained as

$$P_{11}^1(\tau) = Pr\{T_1 < T_2, T_1 < \tau | H1 = H2 = 1\},$$
$$= \int_0^\tau \left( f_{T_1}(t_1) \int_{t_1}^\infty f_{T_2}(t_2) dt_2 \right) dt_1, \quad (7.20)$$

where, $f_{T_1}(t_1)$ and $f_{T_2}(t_2)$ are pdfs under $H_m = 1$ in Eq. (7.14).

Since both the channels are ON, $P_{11}^1(\tau) = P_{11}^2(\tau)$. The probability of interference in this case, i.e., $PI_{11}(\tau)$, can be obtained as

$$PI_{11}(\tau) = 8\left[P_{11}^1(\tau)\right]^3 + 6\left[P_{11}^1(\tau)\right]^2 \left[1 - 2P_{11}^1(\tau)\right]. \quad (7.21)$$

Once the *PS* and the *PI* for all three cases are obtained, $P_{ISO}(\tau)$ and $P_{EIO}(\tau)$ can be obtained as

$$P_{ISO}(\tau) = 1 - (1-p)^2 PS_{00}(\tau) - 2p(1-p)PS_{01}(\tau), \quad (7.22)$$

and $$P_{EIO}(\tau) = 2p(1-p)PI_{01}(\tau) + p^2 PI_{11}(\tau). \quad (7.23)$$

### 7.4.1.1 Performance Using Any Value of *M* with Fixed *L*

In this section, the performance analysis is extended to any value of *M* while keeping $L = 2$, $S_d = L_d = X_d = F_d = 1$ and $I_d = 0$. With $L = 2$, there are three possible cases. When both the channels are OFF, $PS_{00}(\tau)$ can be obtained as

$$PS_{00}(\tau) = \sum_{i=0}^{M-3} 2^{(M-i)} \binom{M}{i} \left[P_{00}^1(\tau)\right]^{(M-i)} \left[1 - 2P_{00}^1(\tau)\right]^i$$
$$+ 2\binom{M}{M-2} \left[P_{00}^1(\tau)\right]^2 \left[1 - 2P_{00}^1(\tau)\right]^{(M-2)}. \quad (7.24)$$

When channel 1 is OFF and 2 is ON, $PS_{01}(\tau)$ and $PI_{01}(\tau)$ can be given by Eq. (7.25) and Eq. (7.26). In these equations, $gcd(a, b)$ represents the greatest common divisor of $a$ and $b$ and $mod(a, b)$ represents the modulo operation which gives the remainder after division of $a$ by $b$.

$$PS_{01}(\tau) = \sum_{i=0}^{M-3} \binom{M}{i} \sum_{j=0}^{\lfloor \frac{M-i}{2} \rfloor - 1} \left[ \binom{M-i}{j} \left[P_{01}^1(\tau)\right]^{M-i-j} \left[P_{01}^2(\tau)\right]^j \left[1 - P_{01}^1(\tau) - P_{01}^2(\tau)\right]^i \right]$$
$$+ \frac{\binom{M-i}{\lfloor \frac{M-i}{2} \rfloor} \left[P_{01}^1(\tau)\right]^{M-i-\lfloor \frac{M-i}{2} \rfloor} \left[P_{01}^1(\tau)\right]^{\lfloor \frac{M-i}{2} \rfloor} \left[1 - P_{01}^1(\tau) - P_{01}^2(\tau)\right]^i}{gcd(i-1,2)mod(M,2) + gcd(i,2)mod(M-1,2)}$$
$$+ \binom{M}{M-2} \left[P_{01}^1(\tau)\right]^2 \left[1 - P_{01}^1(\tau) - P_{01}^2(\tau)\right]^{M-2}. \quad (7.25)$$

$$PI_{01}(\tau) = \sum_{i=0}^{M-3} \binom{M}{i} \sum_{j=0}^{\lfloor \frac{M-i}{2} \rfloor - 1} \left[ \binom{M-i}{j} \left[P_{01}^2(\tau)\right]^{M-i-j} \left[P_{01}^1(\tau)\right]^j \left[1 - P_{01}^2(\tau) - P_{01}^1(\tau)\right]^i \right]$$
$$+ \frac{\binom{M-i}{\lfloor \frac{M-i}{2} \rfloor} \left[P_{01}^2(\tau)\right]^{M-i-\lfloor \frac{M-i}{2} \rfloor} \left[P_{01}^2(\tau)\right]^{\lfloor \frac{M-i}{2} \rfloor} \left[1 - P_{01}^2(\tau) - P_{01}^1(\tau)\right]^i}{gcd(i-1,2)mod(M,2) + gcd(i,2)mod(M-1,2)} +$$
$$\binom{M}{M-2} \left[P_{01}^2(\tau)\right]^2 \left[1 - P_{01}^2(\tau) - P_{01}^1(\tau)\right]^{M-2}. \quad (7.26)$$

# Cooperative Wideband Spectrum Sensing

Finally, when we consider both the channels as ON, the $PI_{11}(\tau)$ can be obtained as

$$PI_{11}(\tau) = \sum_{i=0}^{M-3} 2^{(M-i)} \binom{M}{i} [P_{11}^1(\tau)]^{(M-i)} [1 - 2P_{11}^1(\tau)]^i$$
$$+ 2\binom{M}{M-2} [P_{11}^1(\tau)]^2 [1 - 2P_{11}^1(\tau)]^{(M-2)}. \quad (7.27)$$

Substituting Eq. (7.24) to Eq. (7.27) in Eq. (7.22) and Eq. (7.23) give us $P_{ISO}(\tau)$ and $P_{EIO}(\tau)$.

### 7.4.1.2 Performance Using Any Value of L with Fixed M

In this section, the expressions are derived for $P_{ISO}$ and $P_{EIO}$ for any value of $L$ by considering $M = 3$, $S_d = L_d = X_d = F_d = 1$ and $I_d = 0$. To do this, let us denote pdfs in Eq. (7.14) under $H_m = 0$ and $H_m = 1$ as $f_{H0}(x)$ and $f_{H1}(x)$, respectively. The respective complementary cumulative distribution functions (ccdfs) are denoted by $\bar{F}_{H0}(x)$ and $\bar{F}_{H1}(x)$ which can be obtained by integrating the pdfs from $x$ to $\infty$. Using these, one can obtain $P_{ISO}(\tau)$ and $P_{EIO}(\tau)$ as

$$P_{ISO}(\tau) = 1 - \sum_{i=0}^{L-1} \binom{L}{i}(1-p)^{L-i}p^i \Big[((L-i) + 3(L-i)(L-i-1))I_1^3 +$$
$$3i(L-i)I_1^2 I_2 + 3(L-i)I_1^2(1-(L-i)I_1 - iI_2)\Big]. \quad (7.28)$$

$$P_{EIO}(\tau) = \sum_{i=1}^{L} \binom{L}{i}(1-p)^{L-i}p^i \Big[(i + 3i(i-1))I_2^3$$
$$+ 3i(L-i)I_2^2 I_1 + 3iI_2^2(1-(L-i)I_1 - iI_2)\Big]. \quad (7.29)$$

where, $I_1$ and $I_2$ are given by the following integrals

$$I_1 = \int_0^\tau f_{H0}(x) [\bar{F}_{H0}(x)]^{L-i-1} [\bar{F}_{H_1}(x)]^i dx. \quad (7.30)$$

$$I_2 = \int_0^\tau f_{H1}(x) [\bar{F}_{H1}(x)]^{i-1} [\bar{F}_{H_0}(x)]^{L-i} dx. \quad (7.31)$$

### 7.4.2 THEORETICAL ANALYSIS FOR ALGORITHM 5

Although the approximation given in this chapter is better at approximating the actual pdf compared to the approximation used in Chapter 6, it is mathematically involved to extend the pdf to more than two diversity branches and $M$ cooperating secondary users in case of CSS. Also, the approximation given in Chapter 6 (Section 6.3) works well for values of SNR below 0 $dB$. Hence, in this section we make use of the approximated pdf derived in Chapter 6 (Section 6.3) for SLC. In this case, the approximate probability density function (pdf) for received energy, i.e., decision statistic, under SLC diversity with $P$ diversity branches at the CSUs Eq. (7.6) is given by [refer to Section 6.3, Eq. (6.13)],

$$f_{T_m}(t) \approx \begin{cases} \frac{u(t)a^P \Gamma(1+q)^P t^{Pq+P-1} e^{-\alpha t}}{\Gamma(Pq+P)}, & H_m = 1 \\ \chi^2_{PN}, & H_m = 0 \end{cases} \qquad (7.32)$$

where, $G = N/2$, $D = (1+\bar{\gamma}/m_1)^{m_1}$, $E = \frac{1}{2} - \frac{m_1}{2(m_1+\bar{\gamma})}$, $a = \frac{1}{2^G D \Gamma(G)}$, $a_1 = \frac{(2m_1 E - G)}{2^{G+1} D \Gamma(G+1)}$, $\alpha = -\frac{a_1}{a}$, $q = G - 1$, $m_1$ is the Nakagami parameter, $u(t)$ represents unit step function, $\bar{\gamma}$ is the average signal to noise ratio and $\Gamma(x)$ represents the complete Gamma function.

The received decision statistic for all the channels from all the CSUs are added at the FC in order to obtain new decision statistic as given in Eq. (7.8). Hence, the pdf of this new decision statistic at the FC can be obtained by convolving the pdfs given in Eq. (7.32) with themselves. After carrying out the mathematical simplifications, the pdf of decision statistic at the fusion center is obtained as

$$f_{T_m}(t) \approx \begin{cases} \frac{u(t)a^{MP} \Gamma(1+q)^{MP} t^{MPq+MP-1} e^{-\alpha t}}{\Gamma(MPq+MP)}, & H_m = 1 \\ \chi^2_{MPN}, & H_m = 0 \end{cases} \qquad (7.33)$$

The analysis is now carried out using Eq. (7.33) in this section. To make the analysis easy to understand, the expressions are derived for $P_{EIO}$ and $P_{ISO}$ for $L = 2$, $S_d = L_d = 1$ and $I_d = 0$. With $L_d = 1$, the FC takes decision on the channel which is appearing first after arranging them in ascending order. Since $L = 2$, there are three possible cases for the occupancy of the channels, i.e., both the channels are OFF, any one channel is OFF and both the channels are ON. To simplify the notation, we denote $T_{m,css}$ as $T_m$ for rest of the paper.

Let us start with the case when both the channels are OFF, i.e., $H_1 = 0, H_2 = 0$. In this case, the probability of spectrum opportunity, i.e., $PS_{00}(\tau)$, can be obtained as

$$PS_{00}(\tau) = Pr\{T_1 < T_2, T_1 < \tau | H_1 = H_2 = 0\} + Pr\{T_2 < T_1, T_2 < \tau | H_1 = H_2 = 0\},$$

$$= \int_0^\tau \left[ f_{T_1}(t_1) \int_{t_1}^\infty f_{T_2}(t_2) dt_2 \right] dt_1 + \int_0^\tau \left[ f_{T_2}(t_2) \int_{t_2}^\infty f_{T_1}(t_1) dt_1 \right] dt_2,$$

$$= \int_0^\tau f_{T_1}(t_1) \bar{F}_{T_2}(t_1) dt_1 + \int_0^\tau f_{T_2}(t_2) \bar{F}_{T_1}(t_2) dt_2. \qquad (7.34)$$

As discussed earlier, $f_{T_1}(t_1)$ and $f_{T_2}(t_2)$ are the pdfs of decision statistic obtained at the FC for channel 1 and 2, respectively, which are given by Eq. (7.8) under $H_m = 0$.

Here, $\bar{F}_{T_1}(t_1)$ and $\bar{F}_{T_2}(t_2)$ represent the complementary cumulative distribution function (ccdf) of $f_{T_1}(t_1)$ and $f_{T_2}(t_2)$, respectively. To simplify the notation, let us denote pdf under $H_m = 0$ in Eq. (7.8) as $f_{H0}(t)$ and the corresponding ccdf as $\bar{F}_{H0}(t)$. Since both the channels are OFF, we get $f_{T_1}(t_1) = f_{T_2}(t_2) = f_{H0}(t)$. Using this, the $PS_{00}(\tau)$ can be written as

$$PS_{00}(\tau) = 2\int_0^\tau f_{H0}(t)\bar{F}_{H0}(t)dt. \tag{7.35}$$

The $\bar{F}_{H0}(t)$ can be obtained by integrating pdf given in Eq. (7.8) under $H_m = 0$ form $t$ to $\infty$. After carrying out the mathematical simplification, $\bar{F}_{H0}(t)$ can be obtained as

$$\bar{F}_{H0}(t) = \frac{\Gamma\left(MPG, \frac{t}{2}\right)}{\Gamma(MPG)}, \tag{7.36}$$

where, $\Gamma(a,t)$ represents the upper incomplete Gamma function. Since both the channels are OFF, there is no chance of interference and hence the probability of interference is zero, i.e., $PI(\tau|H_1 = 0, H_2 = 0) = PI_{00}(\tau) = 0$.

We now consider the case when both the channels are ON. In this case, spectrum opportunity does not exist and hence $PS(\tau|H_1 = 1, H_2 = 1) = PS_{11}(\tau) = 0$. The probability of interference opportunity, i.e., $PI_{11}(\tau)$, under $H_1 = H_2 = 1$ can be obtained as

$$PI_{11}(\tau) = \int_0^\tau \left[ f_{T_1}(t_1) \int_{t_1}^\infty f_{T_2}(t_2)dt_2 \right] dt_1, \tag{7.37}$$

where, $f_{T_1}(t_1)$ and $f_{T_2}(t_2)$ are the pdfs of decision statistic given in Eq. (7.8) under $H_m = 1$. Let us denote pdf under $H_m = 1$ in Eq. (7.8) as $f_{H1}(t)$. Since both the channels are ON, we get $f_{T_1}(t_1) = f_{T_2}(t_2) = f_{H1}(t)$ and hence we can write $PI_{11}(\tau)$ as

$$PI_{11}(\tau) = 2\int_0^\tau f_{H1}(t)\bar{F}_{H1}(t)dt, \tag{7.38}$$

where, $\bar{F}_{H1}(t)$ is the ccdf of $f_{H1}(t)$ which can be obtained by integrating $f_{H1}(t)$ from $t$ to $\infty$. After carrying out the integration, we get $\bar{F}_{H1}(t)$ as

$$\bar{F}_{H1}(t) = \frac{\Gamma(1+q)^{MP}\Gamma(MPq+MP, \alpha t)\alpha^{MP}}{\Gamma(MPq+MP)\alpha^{MPq+MP}}. \tag{7.39}$$

Finally, if only one channel is ON (assuming channel 1 is OFF and channel 2 is ON), $PS_{01}(\tau)$ can be obtained by calculating the probability of event $\{T_1 < T_2, T_1 < \tau\}$ under $H_1 = 0$ and $H_2 = 1$, i.e.,

$$PS_{01}(\tau) = \int_0^\tau f_{H0}(t)\bar{F}_{H1}(t)dt. \tag{7.40}$$

The probability of interference in this case can be obtained as

$$PI_{01}(\tau) = \int_0^\tau f_{H1}(t)\bar{F}_{H0}(t)dt. \qquad (7.41)$$

Note that, if we consider channel 1 as ON and channel 2 as OFF, we obtain the same results as in Eq. (7.40) and Eq. (7.41), respectively, i.e., $PS_{01}(\tau) = PS_{10}(\tau)$ and $PI_{01}(\tau) = PI_{10}(\tau)$.

Once, the probability of spectrum opportunity and interference are available for all three cases, $P_{ISO}$ and $P_{EIO}$ can be obtained as

$$P_{ISO}(\tau) = 1 - (1-p)^2 PS_{00}(\tau) - 2p(1-p)PS_{01}(\tau), \qquad (7.42)$$

$$P_{EIO}(\tau) = 2p(1-p)PI_{01}(\tau) + p^2 PI_{11}(\tau). \qquad (7.43)$$

This analysis can be extended to general case of $L$ number of channels in the partial band with $S_d = L_d = 1$ and $I_d = 0$ as

$$P_{ISO}(\tau) = 1 - \left[ \sum_{i=0}^{L-1} \binom{L}{i}(1-p)^{L-i}p^i(L-i) \times \int_0^\tau f_{H0}(t)\left(\bar{F}_{H0}(t)\right)^{L-i-1}\left(\bar{F}_{H1}(t)\right)^i dt \right]. \qquad (7.44)$$

$$P_{EIO}(\tau) = \sum_{i=1}^{L} \binom{L}{i}(1-p)^{L-i}p^i\, i \times \int_0^\tau f_{H1}(t)\left(\bar{F}_{H1}(t)\right)^{i-1}\left(\bar{F}_{H0}(t)\right)^{L-i} dt. \qquad (7.45)$$

## 7.5 RESULTS AND DISCUSSION

In this section, we carry out the expressions using the theoretical analysis given in Section 7.4 for Algorithms 4 and 5. The performance is illustrated using the plots of $P_{EIO}$ Vs. $P_{ISO}$. This section is divided into two subsections to carry out the analysis for the algorithms discussed in Section 7.2. The theoretical analysis is validated using Monte Carlo simulations for both the algorithms. The effect of different parameters on the performance of the algorithms is also illustrated using plots. Such an analysis is essential for any algorithm to optimally select the parameters to achieve better detection performance.

### 7.5.1 EXPERIMENTATIONS USING ALGORITHM 4

In this section, we analyze the Algorithm 4 discussed in Section 7.2.1. The performance is illustrated using the plots of $P_{EIO}$ Vs $P_{ISO}$ under Nakagami fading channel. In Fig. 7.4, we show the plots obtained using both Monte Carlo simulations and theoretical analysis to validate the expressions derived in Section 7.4.1. The plots shown in Fig. 7.5, Fig. 7.6, and Fig. 7.7 are obtained using theoretical analysis given in

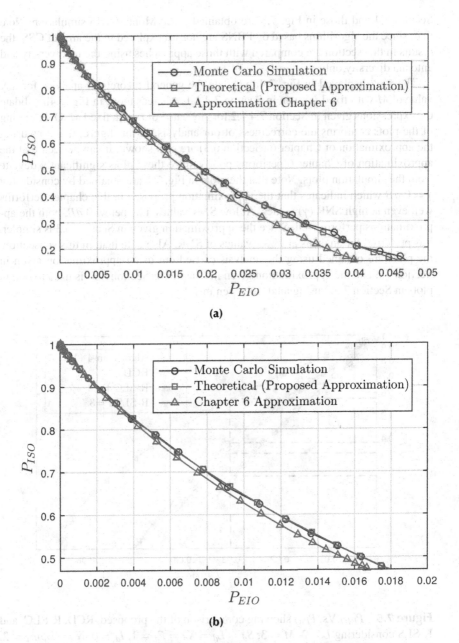

**Figure 7.4** $P_{EIO}$ Vs. $P_{ISO}$ for the proposed algorithm using theoretical analysis and simulation considering $S_d = L_d = X_d = F_d = 1, I_d = 0, N = 10, m_1 = 2, \bar{\gamma} = 0\ dB$ (a) theoretical analysis in Section 7.4.1.1 with $M = 3$ and $L = 2$, (b) theoretical analysis in Section 7.4.1.2 with $L = 4$ and $M = 3$.

Section 7.4 and those in Fig. 7.8 are obtained using Monte Carlo simulation. Note that, since the algorithms based on PBNS are not yet explored in the area of CSS, the results in this section are compared with those approaches using the no diversity and antenna diversity only.

The plot displayed in Fig. 7.4a validate the general theoretical analysis for any value of $M$ with fixed $L$ given in Section 7.4.1.1, whereas those in Fig. 7.4b validate the expressions given in Section 7.4.1.2 for any value of $L$ with fixed $M$. Overlapping of the plots conforms the correctness of our analysis. In this figure, the plots using the approximation of Chapter 6 (Section 6.3) are also shown. It can be seen that the approximation of Chapter 6 performs poorly since these plots significantly deviate from the simulation plots. Note that the plots in Fig. 7.4 are obtained by considering $\bar{\gamma} = 0 \ dB$ which indicates that the approximation proposed in this chapter performs well even at high SNR ($\bar{\gamma}$) values. At low SNR values, i.e., below 0 $dB$, both the approximations perform well. Since the approximation given in Section 6.3 is simpler, it is preferred for analysis at lower values of SNR. Also note that, in this subsection, the plots are obtained using the analysis carried out using approximation given in Section 7.3 and the approximation given in Chapter 6 (Section 6.3) is used to obtain plots in Section 7.5.2 using analysis given in 7.4.2.

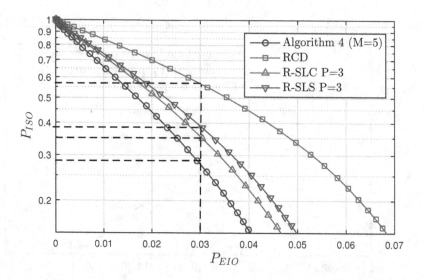

**Figure 7.5** $P_{EIO}$ Vs. $P_{ISO}$ showing comparison of the proposed, RCD, R-SLC, and R-SLS considering $L = 2$, $M = 3$, $S_d = L_d = X_d = F_d = 1$, $I_d = 0$, $N = 10$, $m_1 = 2$, and $\bar{\gamma} = 0 \ dB$.

In Fig. 7.5, the performance of the proposed algorithm is compared against RCD, R-SLC and R-SLS for $M = 3$. The values of $P_{ISO}$ obtained for $P_{EIO} = 0.03$ are tabulated in TABLE 7.1. It can be clearly observed that the proposed algorithm out

### Table 7.1
**Comparison of Algorithm 4 with Other Algorithms**

| Values of $P_{ISO}$ for different algorithms with fixed $P_{EIO} = 0.03$ | | | | |
|---|---|---|---|---|
| Algorithms | Proposed | RCD | R-SLS | R-SLC |
| $P_{ISO}$ | 0.2857 | 0.5686 | 0.3839 | 0.3498 |

### Table 7.2
**Effect of Increasing $M$ on Algorithm 4**

| Values of $P_{ISO}$ for different values of $M$ with fixed $P_{EIO} = 0.03$ | | | | |
|---|---|---|---|---|
| $M$ | 4 | 8 | 12 | 16 |
| $P_{ISO}$ | 0.288 | 0.2289 | 0.1949 | 0.1585 |

performs the RCD. For example, for $P_{EIO} = 0.03$ (see the dotted lines in Fig. 7.5) the $P_{ISO}$ for the proposed algorithm is 50.347% smaller than that of RCD. Also, the proposed algorithm performs better than both R-SLC and R-SLS algorithms in which three diversity branches ($P = 3$) are used. For example, for the same $P_{EIO}$ we get $P_{ISO}$ as 0.2857 for the proposed algorithm and 0.3498 and 0.3839 for R-SLC and R-SLS, respectively. It can be seen that $P_{ISO}$ using the proposed algorithm is 18.325% and 25.579% smaller than R-SLC and R-SLS, respectively. This shows that significant performance improvement can be achieved using our method just by using three CSUs only. It may be mentioned here that the methods proposed in [16, 155] also use the CSS for wide band sensing. However, unlike our approach, they make use of soft combining and the sampling is done considering the entire wide band. Because of this, their approach might perform better than ours. But this happens at the cost of additional bandwidth for transmitting control signals and requiring higher sampling rate. Hence, we refrain from comparing performance of our algorithm with those using soft combining as proposed in [16, 155]. Note that, for antenna diversity the spacing between antennas has to be sufficiently high in order to receive independent observations. Due to this, the implementation of both R-SLC and R-SLS is difficult when the number of diversity branches increases. One can see that, the proposed algorithm discussed here performs better than both R-SLC and R-SLS and we need not worry about the antenna spacing. Here, we have used only two diversity branches. However, one may use more diversity branches at the CSUs to improve the performance.

In Fig. 7.6, we demonstrate the effect of increasing the number of CSUs, i.e., $M$, on the performance where, we vary $M$ between 4 and 16. The values of $P_{ISO}$ for

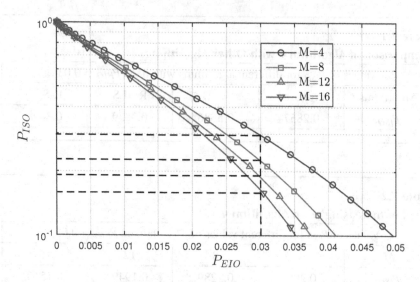

**Figure 7.6** $P_{EIO}$ Vs $P_{ISO}$ using $N = 10$, $S_d = L_d = X_d = F_d = 1$, $I_d = 0$, $p = 0.1$, $m_1 = 2$, $\gamma = 0\ dB$, $L = 2$ for different $M$.

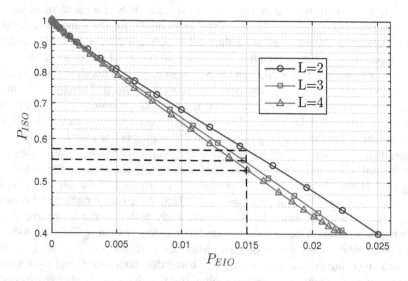

**Figure 7.7** $P_{EIO}$ Vs $P_{ISO}$ using $N = 10$, $S_d = L_d = X_d = F_d = 1$, $I_d = 0$, $p = 0.1$, $m_1 = 2$, $\gamma = 0\ dB$, $M = 3$ for different $L$.

## Table 7.3
### Effect of Increasing $L$ on Algorithm 4

| Values of $P_{ISO}$ for different values of $L$ with fixed $P_{EIO} = 0.015$ | | | |
|---|---|---|---|
| $L$ | 2 | 3 | 4 |
| $P_{ISO}$ | 0.5832 | 0.5498 | 0.5275 |

## Table 7.4
### Effect of Increasing $F_d$ on Algorithm 4

| Values of $P_{ISO}$ for different values of $F_d$ with fixed $P_{EIO} = 0.07$ | | | |
|---|---|---|---|
| $F_d$ | 1 | 2 | 3 |
| $P_{ISO}$ | 0.1274 | 0.0823 | 0.0478 |

$P_{EIO} = 0.03$ is tabulated in TABLE 7.2. We see that for a value of $P_{EIO} = 0.03$, the values of $P_{ISO}$ with $M = 4$ and $M = 16$ are 0.288 and 0.1585, respectively, indicating that the $P_{ISO}$ with $M = 16$ is 44.96% smaller than that with $M = 4$. This shows that the performance can be significantly improved by increasing the number of CSUs. This indicates that, instead of increasing the number of diversity branches, one may increase the number of CSUs to obtain better performance. The effect of increasing number of channels in the partial band, i.e., $L$, is illustrated in Fig. 7.7. The values of $P_{ISO}$ for $P_{EIO} = 0.015$ is tabulated in TABLE 7.3. We observe that with increase in $L$, the performance improves. This is because, increasing $L$ results in more number of free channels in the partial band. One can see that the performance improvement is not significant when $L$ is increased beyond 3. From this we may conclude that after certain value of $L$, the improvement tends to saturate and one does not get an advantage by simply increasing $L$.

Finally, in Fig. 7.8, we demonstrate the effect of increasing $F_d$ on the performance. We know that in order to declare a channel as free, it must be reported as free at least $F_d + 1$ times. This indicates that with the increase in $F_d$, we are providing more security against the interference to the PU. In Fig. 7.8, we can clearly see the improvement in the performance with the increase in $F_d$. The values of $P_{ISO}$ for $P_{EIO} = 0.07$ is tabulated in TABLE 7.8. We see that for a value of $P_{EIO} = 0.07$, the values of $P_{ISO}$ with $F_d = 1$ and $F_d = 3$ are 0.1274 and 0.0427, respectively, indicating that the $P_{ISO}$ with $F_d = 3$ is 66.48% smaller than that with $F_d = 1$. One has to choose the value of $F_d$ appropriately because it also decides the maximum possible value for probability of spectrum opportunity. If the number of CSUs are more, one may select high value for $F_d$.

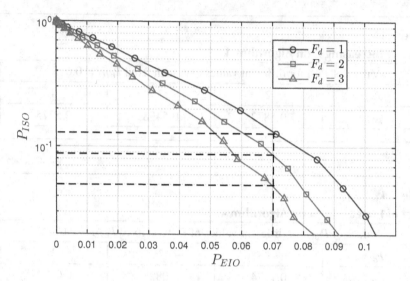

**Figure 7.8** $P_{EIO}$ Vs $P_{ISO}$ using $N = 10$, $S_d = 1$, $I_d = 0$, $p = 0.1$, $\gamma = 0\ dB$, $X_d = 2$, $L = 4$, $M = 8$, $L_d = 2$, $m_1 = 2$ for different $F_d$.

One can also show the effects of other parameters such as $p$, $\gamma$, $L_d$ and $X_d$. However, we refrain from doing it here since the expected results are obvious. For example, increase in the occupancy probability $p$ results in the poor performance due to more number of ON channels falling into the partial band. Hence, we choose not to give these plots here.

### 7.5.2 EXPERIMENTATIONS USING ALGORITHM 5

In this section, we demonstrate and discuss simulation results obtained using theoretical analysis given in Section 7.4.2. As used earlier for Algorithm 4, the performance is illustrated using the plots of $P_{EIO}$ Vs $P_{ISO}$ under Nakagami fading channel.

In Fig. 7.9, we plot $P_{EIO}$ Vs. $P_{ISO}$ using theoretical analysis carried out in Section 7.4.2 and using Monte Carlo simulation. We can see that the plots overlap validating the theoretical analysis. Note that, the expressions given in Eq. (7.32) and Eq. (7.33) for pdfs are the approximate expressions and may not qualify as valid pdf. The small gap in Fig. 7.9 between theoretical and simulation plots is because of the approximation.

Fig. 7.10 shows the plots for the comparison of proposed algorithm against RCD, R-SLC, R-SLS and Algorithm 4. Here, we consider two CSUs, i.e., $M = 2$, having two diversity branches, i.e., $P = 2$. The values of $P_{ISO}$ for $P_{EIO} = 0.03$ are tabulated in TABLE 7.5. We see that the proposed algorithm outperforms all of them. For example, for $P_{EIO} = 0.03$, we observe that $P_{ISO}$ for the proposed Algorithm 5 and RCD are 0.2765 and 0.5788, respectively, indicating that $P_{ISO}$ with Algorithm 5 is

# Cooperative Wideband Spectrum Sensing

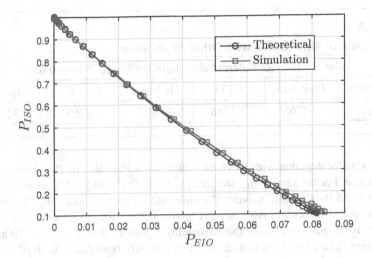

**Figure 7.9** Plots for $P_{EIO}$ Vs. $P_{ISO}$ for the proposed algorithm using theoretical analysis and Monte Carlo simulation considering $M = 2, P = 2, L = 2, S_d = L_d = 1, I_d = 0, m_1 = 2$, and $\bar{\gamma} = -5\ dB$.

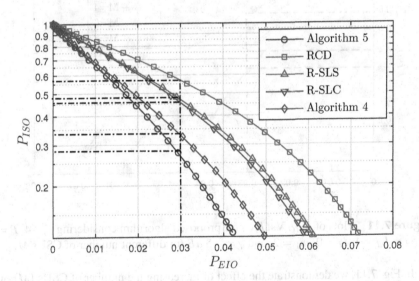

**Figure 7.10** Plots of $P_{EIO}$ Vs. $P_{ISO}$ for the proposed, RCD [185], R-SLS [35], R-SLC [36], and Algorithm 4 considering $M = 2, P = 2, L = 2, S_d = L_d = 1, I_d = 0, m_1 = 2$, and $\bar{\gamma} = 0\ dB$.

**Table 7.5 Comparison of Algorithm 5 with Other Algorithms**

| Values of $P_{ISO}$ for different algorithms with fixed $P_{EIO} = 0.03$ | | | | | |
|---|---|---|---|---|---|
| Algorithm | Proposed | RCD | R-SLS | R-SLC | Algo-4 |
| $P_{ISO}$ | 0.2765 | 0.5788 | 0.4897 | 0.4769 | 0.3231 |

52.22% smaller than that with RCD. Note that lower the value of $P_{ISO}$ better is the performance. For the same $P_{EIO}$, we get $P_{ISO}$ for R-SLC and R-SLS approximately as 0.4769 and 0.4897, respectively, showing inferior performance when compared to the proposed algorithm. Also, we get $P_{ISO} = 0.3231$ for the Algorithm 4 which indicates better performance of the Algorithm 5. This is shown in Fig. 7.10 as the dotted lines. Thus, we see that with only two diversity branches at each of the two CSUs, the detection performance can be significantly improved.

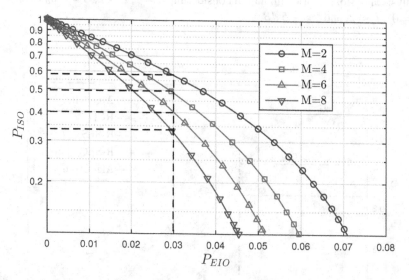

**Figure 7.11** Plots of $P_{EIO}$ Vs. $P_{ISO}$ for proposed algorithm considering $L = 4$, $P = 2$, $S_d = L_d = 1$, $I_d = 0$, $m_1 = 2$, and $\bar{\gamma} = -5\ dB$ for different number of CSUs $M$.

In Fig. 7.11, we demonstrate the effect of increasing the number of CSUs ($M$) on the detection performance. $P_{ISO}$ values for the same $P_{EIO}$ are tabulated in TABLE 7.6. As expected, the detection performance improves significantly. For example, we observe that $P_{ISO}$ for $M = 2$ and $M = 8$ are 0.5789 and 0.3338, respectively, indicating that $P_{ISO}$ with $M = 8$ is 43.33% smaller than that with $M = 2$. We can clearly see significant improvement in performance that by increasing $M$.

### Table 7.6
**Effect on Increasing $M$**

| Values of $P_{ISO}$ for different values of $M$ with fixed $P_{EIO} = 0.03$ | | | | |
|---|---|---|---|---|
| $M$ | 2 | 4 | 6 | 8 |
| $P_{ISO}$ | 0.5787 | 0.4958 | 0.3906 | 0.3338 |

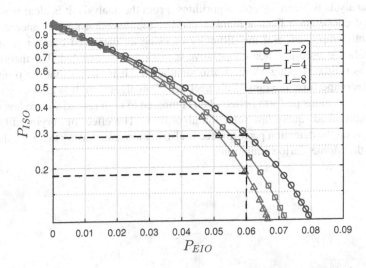

**Figure 7.12** Plots of $P_{EIO}$ Vs. $P_{ISO}$ for proposed algorithm considering $M = 2$, $P = 2$, $S_d = L_d = 1$, $I_d = 0$, $m_1 = 2$, and $\bar{\gamma} = -5$ $dB$ for different values of $L$.

Next, we show in Fig. 7.12 the performance improvement with increase in the number of channels in the partial band ($L$) by varying it from 2 to 8. As seen from the plots, for $P_{EIO} = 0.06$, we get $P_{ISO}$ of 0.2967 and 0.1881 for $L = 2$ and $L = 8$, respectively, which is shown using dotted line in Fig. 7.12 indicating better performance when $L$ is increased.

### Table 7.7
**Effect on Increasing $L$**

| Values of $P_{ISO}$ for different values of $L$ with fixed $P_{EIO} = 0.06$ | | | |
|---|---|---|---|
| $L$ | 2 | 4 | 8 |
| $P_{ISO}$ | 0.2967 | 0.2294 | 0.1881 |

Effects of $p$, $\bar{\gamma}$, $S_d$ and $I_d$ on the performance can also be analyzed. But we choose not to do so since the expected results are obvious. For example, increase in average SNR, i.e., $\bar{\gamma}$, should result in better performance. Hence, we choose not to give these plots.

## 7.6 CONCLUSION

This chapter discussed novel algorithms based on PBNS for cooperative WSS using hard combining and soft combining. A hard combining based algorithm requires lesser control channel bandwidth when compared to soft combining. Complete theoretical analysis is given for both algorithms. From the analysis, it is clear that with the use of cooperation from a number of SUs, the detection performance can be significantly improved, and the difficulty of using an increased number of diversity branches as used in a non-cooperative scenario can be avoided. The theoretical analysis is validated using Monte Carlo simulations. It is shown that the proposed algorithms outperform the ranked channel detection algorithm. It is also shown that proposed algorithms perform better when compared to both ranked square law combining and ranked square law selection algorithms. The effect of varying different parameters on the detection performance is also demonstrated using theoretical analysis and the Monte Carlo simulations.

# 8 Conclusions and Future Research Directions

In this chapter, we provide the conclusions by summarizing our main contributions and also indicate future research directions.

## 8.1 CONCLUSIONS

In this book we have addressed the problem of spectrum sensing for cognitive radio by providing theoretical analysis of energy-detection-based approach considering practical scenarios and by proposing novel detection algorithms. We began with the discussion of energy-detection-based spectrum sensing over $\eta - \lambda - \mu$ fading channel. The $\eta - \lambda - \mu$ fading model represents a general model which includes other existing models as special cases. The performance improvement was shown using diversity and cooperative spectrum sensing. The analysis was then extended to the case of shadowing in addition to fading.

In our next work, we have analyzed the performance of generalized energy detector in the presence of noise uncertainty and fading. First, the expressions for detection probabilities considering AWGN channel were derived for three cases namely, no diversity, with diversity and cooperative spectrum sensing. We then derive the expressions for SNR walls and discussed the effect of diversity and cooperative spectrum sensing on the SNR walls. The analysis was then extended to the channel with Nakagami fading and it was shown that in this scenario, the SNR wall increases significantly. All the derived expressions were validated using Monte Carlo simulations.

Above works were carried out by considering narrowband sensing. Our next work involved detection techniques for wideband spectrum sensing. Two new detection algorithms, namely, ranked square law combining (R-SLC) and ranked square law selection (R-SLS), were proposed that make use of diversity. Complete theoretical analysis was provided for these algorithms considering Nakagami fading. The proposed algorithms outperform the ranked channel detection algorithm when used without diversity. The theoretical analysis was then validated using Monte Carlo simulations. Also, the effects of different parameters on the detection performance was studied.

In our next work, novel detection algorithms for cooperative wideband spectrum sensing were proposed using hard and soft combining. The complete theoretical analysis was provided for the proposed algorithms which are validated using Monte Carlo simulations. It was shown that the proposed hard combining based algorithm outperforms the ranked channel detection algorithm. By choosing appropriate number of cooperating secondary users, the proposed algorithm performs better than both R-SLC and R-SLS. The effects of different parameters on the detection performance

DOI: 10.1201/9781003088554-8

was also studied. It was demonstrated that the soft combining based algorithm outperforms all the other algorithms.

## 8.2 FUTURE RESEARCH DIRECTIONS

This book has presented theoretical analysis of existing narrowband spectrum sensing algorithms considering practical scenarios and few novel detection algorithms for wideband spectrum sensing. In the process of this work, however, we identified related problems that one may consider worth pursuing. These are briefly described as below.

- *Analysis of other detection techniques under generalized fading* :
  In Chapter 4, we have analyzed the performance of energy-detection-based spectrum sensing under general fading model. This analysis can be extended for other spectrum sensing techniques available in literature, namely, matched filter detection, cyclostationary detection, eigenvalue based detection, covariance based detection, etc.
- *Machine learning techniques for spectrum sensing*:
  In Chapters 4 and 5, we addressed the problem of spectrum sensing where it is assumed that primary users transmit at single power level. Recently, a new multiple primary transmit power (MPTP) scenario has been proposed in [113], where PU could work under more than one discrete power levels while the task of spectrum sensing is not only to detect the status of PU, but also to recognize PU's transmit power level. This can be addressed as a classification problem with multiple classes. Machine learning techniques can be applied to solve this problem. The machine learning techniques can also be applied to the traditional scenario where PU transmits at a single power level which can be considered as a problem of binary classification.
- *Analysis of proposed algorithms in practical scenarios*:
  In Chapter 6, we proposed detection algorithms that use SLC and SLS diversity and the analysis was given considering Nakagami fading. One can extend the same to more general system models consisting of varying channel bandwidths, SNR levels across channels, fading, noise uncertainty, etc. The problem can be formulated in an optimization frame work to find the optimal thresholds when different priorities, and different occupancy probabilities are considered for PU channels.
- *Use of soft combining for cooperative wideband spectrum sensing*:
  In Chapter 7, we addressed the problem of cooperative wideband spectrum sensing where we use hard combining at the fusion center. We know that the soft combining gives better performance when compared to hard combining. Hence, the soft combining can be used for solving the problem we considered for cooperative wideband spectrum sensing. However, the use of soft combining requires higher control channel bandwidth. One can use quantized soft combining in order to reduce the required control channel bandwidth.

- *Combining antenna diversity and cooperative spectrum sensing for performance improvement*:
  In all our works where we use cooperative spectrum sensing, we consider that all the cooperating secondary users have only few receive antenna. One can consider the scenario where cooperating secondary users have multiple antennas. In this case, one can get improved performance since it takes advantages of both antenna diversity and CSS. The major problem with this kind of scheme is the computational complexity. One can use quantized cooperative spectrum sensing to reduce the complexity.
- *Cooperative Spectrum Sensing under Imperfect Reporting Channels*:
  In this book, it is assumed that the reporting channels in cooperative spectrum sensing are perfect. In such a scenario, the information transmitted by cooperating secondary users reach the fusion center without any errors. However, in real life, the reporting channels are imperfect and the information transmitted by CSUs gets corrupted due to channel errors. The performance of CSS should be studied under imperfect reporting channels in order to properly understand the performance under real environment. There is also need of effective algorithms that can reduce the effects of reporting channel errors.

# A  Appendix for Chapter 1

## A.1  PROOF FOR MARKOV INEQUALITY

The expected value of random variable is given as

$$E\{X\} = \int_{-\infty}^{+\infty} x f_X(x)\, dx. \tag{A.1}$$

For nonnegative random variable $X$, the expression Eq. (A.1) can be written as

$$E\{X\} = \int_{0}^{+\infty} x f_X(x)\, dx. \tag{A.2}$$

For some positive constant $a$, the integration in Eq. (A.2) can be written as sum of two separate integrals as

$$E\{X\} = \int_{0}^{a} x f_X(x)\, dx + \int_{a}^{+\infty} x f_X(x)\, dx. \tag{A.3}$$

If we discard the integral from 0 to $a$ in Eq. (A.3), we can write the inequality as

$$E\{X\} \geq \int_{a}^{+\infty} x f_X(x)\, dx. \tag{A.4}$$

If we replace $x$ with smaller number $a$, we get the inequality as

$$E\{X\} \geq \int_{a}^{+\infty} a f_X(x)\, dx = a P\{X \geq a\}, \tag{A.5}$$

where,

$$\int_{a}^{+\infty} f_X(x)\, dx = P\{X \geq a\}. \tag{A.6}$$

The inequality given in Eq. (A.5) represents the Markov inequality.

DOI: 10.1201/9781003088554-A

## A.2 PROOF CENTRAL LIMIT THEOREM

In this subsection we present the proof of central limit theorem. We can write $Z_n$ as

$$Z_n = \frac{S_n - n\mu}{\sigma\sqrt{n}} = \frac{1}{\sigma\sqrt{n}} \sum_{i=1}^{n} (X_i - \mu). \tag{A.7}$$

We will derive the characteristic function of $Z_n$ and show that as $n$ approaches infinity, the characteristic function of $Z_n$ approaches that of a Normal random variable. The characteristic function of $Z_n$ using Eq. (1.58) can be obtained as

$$\Phi_{Z_n}(\omega) = E\left\{e^{j\omega Z_n}\right\},$$

$$= E\left\{\exp\left[\frac{j\omega}{\sigma\sqrt{n}} \sum_{i=1}^{n} (X_i - \mu)\right]\right\},$$

$$= E\left\{\prod_{i=1}^{n} \exp\left[\frac{j\omega}{\sigma\sqrt{n}} (X_i - \mu)\right]\right\}. \tag{A.8}$$

The last equality in the above expression results from the independence of $X_i's$. Since expectation operation is linear, we can write the characteristic function of $Z_n$ as

$$\Phi_{Z_n}(\omega) = \prod_{i=1}^{n} E\left\{\exp\left[\frac{j\omega}{\sigma\sqrt{n}} (X_i - \mu)\right]\right\}. \tag{A.9}$$

Since $X_i's$ are identically distributed, we can write

$$\Phi_{Z_n}(\omega) = \left[E\left\{\exp\left[\frac{j\omega}{\sigma\sqrt{n}} (X - \mu)\right]\right\}\right]^n. \tag{A.10}$$

We know the series form expansion of $e^x$ as

$$e^x = \sum_{i=0}^{\infty} \frac{x^i}{i!} = 1 + x + \frac{x^2}{2!} + \frac{x^3}{3!} + \cdots \tag{A.11}$$

We can expand the exponential in Eq. (A.10) using Eq. (A.11) as

$$E\left\{\exp\left[\frac{j\omega}{\sigma\sqrt{n}} (X - \mu)\right]\right\} = E\left\{1 + \frac{j\omega(X-\mu)}{\sigma\sqrt{n}} + \frac{j^2\omega^2(X-\mu)^2}{2!\sigma^2 n} + R(\omega)\right\}$$

$$= 1 + \frac{j\omega}{\sigma\sqrt{n}} E\{(X-\mu)\} + \frac{j^2\omega^2}{2!\sigma^2 n} E\left\{(X-\mu)^2\right\} + E\{R(\omega)\}. \tag{A.12}$$

Note that $E\{(X-\mu)\} = 0$ and $E\left\{(X-\mu)^2\right\} = \sigma^2$. Using this in Eq. (A.12) we have

$$E\left\{\exp\left[\frac{j\omega}{\sigma\sqrt{n}} (X - \mu)\right]\right\} = 1 - \frac{\omega^2}{2n} + E\{R(\omega)\}. \tag{A.13}$$

Appendix for Chapter 1

The term $E\{R(\omega)\}$ can be neglected relative to $\omega^2/2n$ as $n$ becomes large. Using Eq. (A.13) in Eq. (A.10), we get

$$\Phi_{Z_n}(\omega) = \left[1 - \frac{\omega^2}{2n}\right]^2. \tag{A.14}$$

If we let $n \to \infty$ in Eq. (A.14), we get

$$\Phi_{Z_n}(\omega) = e^{\omega^2/2} \text{ as } n \to \infty. \tag{A.15}$$

The right-hand side of the Eq. (A.15) represents the characteristic function of a zero mean, unit variance Gaussian random variable. The derivation for the characteristic function of zero mean, unit variance Gaussian random variable is given in APPENDIX A.3. Hence, we say that the cdf of $Z_n$ approaches cdf of a zero mean and a unit variance Gaussian random variable.

## A.3 CHARACTERISTIC FUNCTION OF GAUSSIAN RANDOM VARIABLE

The probability density function of a zero mean and a unit variance Gaussian random variable is given as

$$f_X(x) = \frac{1}{\sqrt{2\pi}} e^{-x^2/2}. \tag{A.16}$$

The characteristic function of $X$ is obtained as

$$\Phi_X(\omega) = E\{e^{j\omega X}\}. \tag{A.17}$$

The expected value of function of random variable $Y = g(X)$ is obtained as

$$E\{g(X)\} = \int_{-\infty}^{+\infty} g(x) f_X(x) dx. \tag{A.18}$$

We see that Eq. (A.17) is of the form $E\{g(X)\}$ with $g(X) = e^{j\omega X}$. We can make use of Eq. (A.18) to obtain the characteristic function as

$$\Phi_X(\omega) = \int_{-\infty}^{+\infty} e^{j\omega x} f_X(x) dx. \tag{A.19}$$

Substituting Eq. (A.16) in Eq. (A.19), we get

$$\Phi_X(\omega) = \frac{1}{\sqrt{2\pi}} \int_{-\infty}^{+\infty} e^{j\omega x} e^{-\frac{x^2}{2}} dx,$$

$$= \frac{1}{\sqrt{2\pi}} \int_{-\infty}^{+\infty} e^{-\left(\frac{x^2}{2} - j\omega x\right)} dx,$$

$$= \frac{1}{\sqrt{2\pi}} \int_{-\infty}^{+\infty} e^{-\frac{1}{2}\left(x^2 - 2j\omega x\right)} dx,$$

$$= \frac{1}{\sqrt{2\pi}} \int_{-\infty}^{+\infty} e^{-\frac{1}{2}\left(x^2 - 2j\omega x + j^2\omega^2 - j^2\omega^2\right)} dx,$$

$$= \frac{1}{\sqrt{2\pi}} \int_{-\infty}^{+\infty} e^{-\frac{1}{2}(x-jw)^2} e^{\frac{1}{2}j^2\omega^2} dx,$$

$$= e^{-\frac{\omega^2}{2}} \int_{-\infty}^{+\infty} \underbrace{\frac{1}{\sqrt{2\pi}} e^{-\frac{1}{2}(x-jw)^2}}\, dx. \tag{A.20}$$

The expression marked with under braces in above equation represents the pdf of Gaussian distributed random with mean $j\omega$ and unit variance. The pdf when integrated over the interval $-\infty$ to $\infty$ gives the area under the pdf which is always equals to one. Using this in above equation, we get the characteristic function of Gaussian random variable as

$$\Phi_X(\omega) = e^{-\frac{\omega^2}{2}}. \tag{A.21}$$

# B Appendix for Chapter 4

## B.1 DERIVATION FOR $P_F(\tau)$ IN EQ. (4.3)

The probability of false alarm in Eq. (4.3) is given by $P_F = Pr\{T_m > \tau | H_0\}$ and is derived as follows.

$$P_F(\tau) = \int_\tau^\infty f_{T_m}(t) dt = \int_\tau^\infty \chi_N^2 \, dt, \tag{B.1}$$

where, $f_{T_m}(t)$ is the pdf of decision statistic $T_m$ under hypotheses $H_0$ and is given in Eq. (4.2). The pdf of chi-square distribution is given by

$$\chi_N^2 = \frac{1}{2^{\frac{N}{2}} \Gamma\left(\frac{N}{2}\right)} t^{\frac{N}{2}-1} e^{-\frac{t}{2}}. \tag{B.2}$$

Substituting pdf from Eq. (B.2) into Eq. (B.1) we get

$$P_F(\tau) = \int_\tau^\infty \frac{1}{2^{\frac{N}{2}} \Gamma\left(\frac{N}{2}\right)} t^{\frac{N}{2}-1} e^{-\frac{t}{2}} dt. \tag{B.3}$$

Using the change of variable as $\frac{t}{2} = x$ and some mathematical simplifications, we can write Eq. (B.3) as

$$P_F(\tau) = \frac{1}{\Gamma\left(\frac{N}{2}\right)} \int_{\frac{\tau}{2}}^\infty x^{\frac{N}{2}-1} e^{-x} dx = \frac{\Gamma\left(\frac{N}{2}, \frac{\tau}{2}\right)}{\Gamma\left(\frac{N}{2}\right)}. \tag{B.4}$$

# C Appendix for Chapter 5

## C.1 DERIVATION FOR $\bar{P}_{D,PLC}$ IN EQ. (5.27)

To derive $\bar{P}_{D,plc}$, we require the following result

$$\lim_{N\to\infty} Q\left(a\sqrt{N}\right) = \begin{cases} 0, & \text{if } a > 0, \\ 1, & \text{if } a < 0, \\ 0.5 & \text{if } a = 0. \end{cases} \quad \text{(C.1)}$$

Applying limit $N \to \infty$ to Eq. (5.25) and using Eq. (C.1), we get

$$\bar{P}_{D,plc} = \int_{a_1}^{b_1}\int_{a_2}^{b_2} \mathcal{H}\left[G_p\left((1+\tilde{\gamma}_1 x)^{\frac{p}{2}} + (1+\tilde{\gamma}_2 y)^{\frac{p}{2}}\right) - 2\lambda(xy)^{\frac{p}{2}}\right] \frac{25}{L_1 L_2 (\ln(10))^2 xy} dy dx, \quad \text{(C.2)}$$

where, $\mathcal{H}(x)$ represents the Heaviside function defined as

$$\mathcal{H}(x) = \begin{cases} 0, & \text{if } x < 0, \\ \frac{1}{2}, & \text{if } x = 0, \\ 1 & \text{if } x > 0. \end{cases} \quad \text{(C.3)}$$

Applying the change of variables as $z = \ln(x)$ and $w = \ln(w)$, we get

$$\bar{P}_{D,plc} = C_1 \int_{-c}^{c}\int_{-d}^{d} \mathcal{H}\left[G_p\left((1+e^z\tilde{\gamma}_1)^{\frac{p}{2}} + (1+e^w\tilde{\gamma}_2)^{\frac{p}{2}}\right) - 2\lambda e^{\frac{zp}{2}} e^{\frac{wp}{2}}\right] dw dz, \quad \text{(C.4)}$$

where, $c = \ln(b_1)$, $d = \ln(b_2)$ and $C_1 = 100/L_1 L_2 p^2 \ln(10)^2$.

We will evaluate this integral for three ranges of $\lambda$. First, we evaluate $\bar{P}_{D,plc}$ for

$$\lambda \leq \frac{G_p}{2}\left\{\left(\tilde{\gamma}_1 + e^{-c}\right)^{\frac{p}{2}} + \left(\tilde{\gamma}_2 + e^{-d}\right)^{\frac{p}{2}}\right\}. \quad \text{(C.5)}$$

In this range, the argument of $\mathcal{H}[\cdot]$ in Eq. (C.4) is $> 0$ for any $z < c$ or $w < d$. Hence, we get $\bar{P}_{D,plc}$ as

$$\bar{P}_{D,plc} = C_1 \int_{-c}^{c}\int_{-d}^{d} (1) dw dz = 1. \quad \text{(C.6)}$$

The next range of $\lambda$ is considered as

$$\lambda \geq \frac{G_p}{2}\left\{(\tilde{\gamma}_1+e^c)^{\frac{p}{2}}+(\tilde{\gamma}_2+e^d)^{\frac{p}{2}}\right\}. \qquad (C.7)$$

In this range, the argument of $\mathcal{H}[\cdot]$ in Eq. (C.4) is $<0$ for $z>-c$ or $w>-d$. Hence, we get $\bar{P}_{D,plc}$ as

$$\bar{P}_{D,plc} = C_1 \int_{-c}^{c}\int_{-d}^{d}(0)dwdz = 0. \qquad (C.8)$$

Finally, we consider the range of $\lambda$ as

$$\frac{G_p}{2}\left\{(\tilde{\gamma}_1+e^{-c})^{\frac{p}{2}}+(\tilde{\gamma}_2+e^{-d})^{\frac{p}{2}}\right\} < \lambda < \frac{G_p}{2}\left\{(\tilde{\gamma}_1+e^c)^{\frac{p}{2}}+(\tilde{\gamma}_2+e^d)^{\frac{p}{2}}\right\}. \qquad (C.9)$$

Here, we can find the integration range of $w$ in terms of $z$ for which the argument of $\mathcal{H}[\cdot]$ is $>0$. After some mathematical simplifications, we get this condition as

$$w < -\ln\left[\left\{A-(e^{-z}+\tilde{\gamma}_1)^{\frac{p}{2}}\right\}^{\frac{2}{p}} - \tilde{\gamma}_2\right], \qquad (C.10)$$

where, $A = 2\lambda/G_p$. This will be the upper limit of $w$ for which $\mathcal{H}(\cdot)$ in Eq. (C.4) becomes 1. Note that, we need to make sure that the upper limit does not go beyond $d$ and below $-d$. Hence, the upper limit for $w$ can be written as

$$U_1 = \max\left[\min\left\{d, -\ln\left[\left\{A-(e^{-z}+\tilde{\gamma}_1)^{\frac{p}{2}}\right\}^{\frac{2}{p}} - \tilde{\gamma}_2\right]\right\}, -d\right]. \qquad (C.11)$$

Using this, we can write $\bar{P}_{D,plc}$ as

$$\bar{P}_{D,plc} = C_1 \int_{-c}^{c}\int_{-d}^{U_1}(1)dwdz = C_1\underbrace{\int_{-c}^{c}U_1 dz}_{I_1} + 2cd. \qquad (C.12)$$

The integration limit of $I_1$ in Eq. (C.12) can be split into two parts as

$$I_1 = \underbrace{\int_{-c}^{U_2}(d)dz}_{I_{11}} + \underbrace{\int_{U_2}^{c}\max[R_2(-z),-d]dz}_{I_{12}} \qquad (C.13)$$

where, $R_2(z) = -\ln\left(\left(A-(e^x+\tilde{\gamma}_2)^{\frac{p}{2}}\right)^{\frac{2}{p}} - \tilde{\gamma}_1\right)$ and $U_2 = \max[-c, R_2(-d)]$. The $I_{11}$ in Eq. (C.13) is reduced as

$$I_{11} = dU_2 + dc. \qquad (C.14)$$

# Appendix for Chapter 5

Now, the integration limit of $I_{12}$ in Eq. (C.13) can again split into two parts as

$$I_{12} = \int_{U_2}^{U_3} -\ln\left[\left\{A - (e^{-z} + \tilde{\gamma}_1)^{\frac{p}{2}}\right\}^{\frac{2}{p}} - \tilde{\gamma}_2\right] dz + \int_{U_3}^{c} (-d)\, dz, \quad (C.15)$$

where, $U_3$ is given as $U_3 = \min[c, R_2(d)]$.

It is mathematically too involved to integrate the first integral in Eq. (C.15) and hence we approximate the $\ln(x)$ term by using the approximation $\ln(x+1) \approx x$ for small $x$. Using this approximation, we can write $I_{12}$ as

$$I_{12} \approx -\int_{U_2}^{U_3} \left\{A - (e^{-z} + \tilde{\gamma}_1)^{\frac{p}{2}}\right\}^{\frac{2}{p}} dz + \int_{U_2}^{U_3} \tilde{\gamma}_2 + 1\, dz + \int_{U_3}^{c} (-d)\, dz. \quad (C.16)$$

Since it is mathematically involved to integrate the first integral in Eq. (C.16), we first approximate its integrand using asymptotic analysis. Using the Taylor series expansion of a function $f(t)$, we can write [58]

$$f(z) = aa z^q + aa_1 z^{q+1} + O(z^{q+2}) \text{ as } z \to 0^+. \quad (C.17)$$

Here, $aa$, $aa_1$ and $q$ represent real constants and $O(t^{q+2})$ is the error term as $t \to 0^+$. Using Eq. (C.17), the authors in [58] propose the approximation as

$$f(z) \approx aa z^q e^{-\alpha z}, \text{ as } z \to 0^+, \quad (C.18)$$

where, $\alpha = -\frac{aa_1}{aa}$. The Taylor series expansion of integrand of first integral in Eq. (C.16) can be obtained as

$$f(z) = \left(A - (1 + \tilde{\gamma}_1)^{\frac{p}{2}}\right)^{\frac{2}{p}} + (1 + \tilde{\gamma}_1)^{\frac{p}{2}-1} \left(A - (1 + \tilde{\gamma}_1)^{\frac{p}{2}}\right)^{\frac{2}{p}-1} z + O(z^2), \quad (C.19)$$

as $z \to 0^+$. On comparing Eq. (C.19) with Eq. (C.18) we get $aa = \left(A - (1 + \tilde{\gamma}_1)^{\frac{p}{2}}\right)^{\frac{2}{p}}$, $aa_1 = (1 + \tilde{\gamma}_1)^{\frac{p}{2}-1} \left(A - (1 + \tilde{\gamma}_1)^{\frac{p}{2}}\right)^{\frac{2}{p}-1}$, $\alpha = \frac{-aa_1}{aa}$ and $q = 0$. Using this approximation, we can write $I_{12}$ as

$$I_{12} \approx -\int_{U_2}^{U_3} aa\, e^{-\alpha z} dz + \int_{U_2}^{U_3} (\tilde{\gamma}_2 + 1)\, dz + \int_{U_3}^{c} (-d)\, dz. \quad (C.20)$$

After performing the integration $I_{12}$ can be reduced as

$$I_{12} \approx \frac{aa}{\alpha} \left[e^{-\alpha U_3} - e^{-\alpha U_2}\right] + (\tilde{\gamma}_2 + 1)(U_3 - U_2) - d(c - U_3). \quad (C.21)$$

Substituting $I_{11}$ and $I_{12}$ from Eq. (C.14) and Eq. (C.21), respectively, into Eq. (C.13) gives us the reduced expression for $I_1$. Using this $I_1$ in Eq. (C.12), we get the reduced expression for $\bar{P}_{D,plc}$ for the final range of $\lambda$. This expression combined with Eq. (C.6) and Eq. (C.8) gives us the expression given in Eq. (5.27).

## C.2 DERIVATION FOR $\bar{P}_D^{NAK}$ IN EQ. (5.77)

After applying the change of variable as $w = \ln(x)$ in Eq. (5.74) and applying limit $N \to \infty$, using Eq. (C.3), the $\bar{P}_D^{Nak}$ can be written as

$$\bar{P}_D^{Nak} = \int_0^\infty \int_{\ln(a)}^{\ln(b)} \mathcal{H}\left(G_p(1+\bar{\gamma}e^w z)^{\frac{p}{2}} - \lambda e^{\frac{wp}{2}}\right) \frac{5m^m z^{m-1} e^{-mz}}{L\ln(10)\Gamma(m)} dwdz. \quad (C.22)$$

The range of $w$ in terms of $z$ for which argument of $\mathcal{H} > 0$ can be found as

$$w < -\ln\left[\left(\frac{\lambda}{G_p}\right)^{\frac{2}{p}} - z\bar{\gamma}\right]. \quad (C.23)$$

Using this, we can write $\bar{P}_D^{Nak}$ as

$$\bar{P}_D^{Nak} = \frac{5m^m}{L\ln(10)\Gamma(m)} \int_0^\infty \int_{\ln(a)}^{U_4} \frac{z^{m-1}}{e^{mz}} dwdz = \frac{5}{L\ln(10)}\left[\underbrace{\int_0^\infty U_4 \frac{m^m z^{m-1}}{\Gamma(m)e^{mz}}dz}_{I_1} - \ln(a)\underbrace{\int_0^\infty \frac{m^m z^{m-1}}{\Gamma(m)e^{mz}}dz}_{I_2}\right], \quad (C.24)$$

where, $U_4$ is given as

$$U_4 = \max\left[\min\left[\ln(b), -\ln\left[\left(\frac{\lambda}{G_p}\right)^{\frac{2}{p}} - z\bar{\gamma}\right]\right], \ln(a)\right]. \quad (C.25)$$

The integral $I_1$ in Eq. (C.24), can be written as

$$I_1 = \underbrace{\int_0^{U_5} \frac{\ln(a) m^m z^{m-1}}{\Gamma(m)e^{mz}}dz}_{I_{11}} + \underbrace{\int_{U_5}^{U_6} -\ln\left(\frac{\lambda}{G_p} - z\bar{\gamma}\right) \frac{m^m z^{m-1}}{\Gamma(m)e^{mz}}dz}_{I_{12}} + \underbrace{\int_{U_6}^\infty \frac{\ln(b) m^m z^{m-1}}{\Gamma(m)e^{mz}}dz}_{I_{13}}, \quad (C.26)$$

where, we use $R_3(z) = \left(\frac{1}{\bar{\gamma}}\left(\frac{\lambda}{G_p}\right)^{\frac{2}{p}} - \frac{1}{z\bar{\gamma}}\right)$, $U_5 = \max[0, R_3(a)]$ and $U_6 = \max[0, R_3(b)]$.

The $I_{11}$ and $I_{13}$ in Eq. (C.26) can be reduced as

$$I_{11} = \frac{\ln(a)}{m^m} P(m, m\, U_5) \text{ and } I_{13} = \frac{\ln(b)}{m^m} Q(m, m\, U_6), \quad (C.27)$$

where, $P(a,z) = \frac{\gamma(a,z)}{\Gamma(a)}$ and $Q(a,z) = \frac{\Gamma(a,z)}{\Gamma(a)}$ are the regularized lower and upper incomplete Gamma functions.

Appendix for Chapter 5    193

Using the approximation $\ln(x+1) = x$ for small $x$, we can derive approximate $I_{12}$ in Eq. (C.26) as

$$I_{12} \approx \left(1 - \left(\frac{\lambda}{G_p}\right)^{\frac{2}{p}}\right)[P(m, m U_6) - P(m, m U_5)] + \bar{\gamma}[P(m+1, m U_6) - P(m+1, m U_5)]. \quad (C.28)$$

Substituting $I_{11}$, $I_{12}$ and $I_{13}$ from Eq. (C.27) and Eq. (C.28) Eq. (C.26) we get $I_1$. The $I_2$ in Eq. (C.24) can be reduced as $I_2 = \ln(a)$. Using $I_1$ and $I_2$ in Eq. (C.24) and carrying out few simplifications, we get $\bar{P}_D^{Nak}$ given in Eq. (5.77).

## C.3 DERIVATION FOR $\bar{P}_{D,PLC}^{NAK}$ IN EQ. (5.81)

Using transformation of random variable as $t = \ln(x)$ and $u = \ln(y)$ in Eq. (5.80) and applying limit $N \to \infty$ we get

$$\bar{P}_{D,plc}^{Nak} = \frac{25 m^{2m}}{L_1 L_2 (\ln(10))^2} \int_0^\infty \int_0^\infty \int_{-c}^c \int_{-d}^d z^{m-1} e^{-mz} w^{m-1} e^{-mw}$$

$$\mathcal{H}\left[G_p\left(\frac{(1+\bar{\gamma}_1 e^t z)^{\frac{p}{2}}}{e^{\frac{-up}{2}}} + \frac{(1+\bar{\gamma}_2 e^u w)^{\frac{p}{2}}}{e^{\frac{-tp}{2}}}\right) - \frac{2\lambda}{e^{\frac{-(t+u)p}{2}}}\right] du\, dt\, dz\, dw. \quad (C.29)$$

Here, we can find the integration limit of $u$ in terms of $t$, $z$ and $w$ for which the argument of $\mathcal{H}(\cdot)$ is $> 0$. After few mathematical simplifications, we get upper limit of $u$ as

$$u < -\ln\left[\left(\frac{2\lambda}{G_p} - (\bar{\gamma}_1 z + e^{-t})^{\frac{p}{2}}\right)^{\frac{2}{p}} - \bar{\gamma}_2 w\right]. \quad (C.30)$$

Once again, we need to make sure that, the upper limit do not go beyond $d$ and below $-d$. Hence, we can write the upper limit of $u$ as

$$U7 = \max\left[\min\left[-\ln\left[\left(\frac{2\lambda}{G_p} - (\bar{\gamma}_1 z + e^{-t})^{\frac{p}{2}}\right)^{\frac{2}{p}} - \bar{\gamma}_2 w, d\right]\right], -d\right]. \quad (C.31)$$

Hence, we can write $\bar{P}_{D,plc}^{Nak}$ in Eq. (C.29) as

$$\bar{P}_{D,plc}^{Nak} = \frac{25 m^{2m}}{L_1 L_2 (\ln(10))^2} \int_0^\infty \int_0^\infty \int_{-c}^c \int_{-d}^{U7} z^{m-1} e^{-mz} w^{m-1} e^{-mw} du\, dt\, dz\, dw. \quad (C.32)$$

After carrying out the integration with respect to $u$, we get the expression for $\bar{P}_{D,plc}^{Nak}$ given in Eq. (5.81). These are the steps for reducing the expression to three integrals from four integrals. We can continue in this fashion to derive the final expression but the final expression will be too lengthy. Hence, we keep the expression that is reduced to three integrals.

# D Appendix for Chapter 6

## D.1 PROOF FOR CONVERGENCE OF PDF OF SLC UNDER NAKAGAMI FADING CHANNEL IN EQ. (6.12)

We denote series in (6.12) as $a_n$.

$$a_n = \frac{e^{-t}t^{2G-1}}{2^{2G}D^2\Gamma(m_1)^2} \sum_{v=0}^{\infty}\sum_{q=0}^{\infty} \frac{\Gamma(m+v)\Gamma(m+q)E^{v+q}t^{v+q}}{v!q!\Gamma(2G+v+q)}. \quad (D.1)$$

We will prove convergence of $a_n$ using comparison test. We will create new series $b_n$ which is larger than the given series and that converge. Using comparison test we can say that since the larger series converge, the smaller series will converge. We use the fact that $\Gamma(2G+v+q) > \Gamma(G+v)\Gamma(G+q)$ for nonzero positive values of $G$ to create new series. The series $b_n$ is

$$b_n = \sum_{v=0}^{\infty}\sum_{q=0}^{\infty} \frac{\Gamma(m+v)\Gamma(m+q)E^{v+q}t^{v+q}}{v!q!\Gamma(G+v)\Gamma(G+q)}$$

$$= \left(\sum_{v=0}^{\infty}\frac{\Gamma(m+v)E^{v}t^{v}}{v!\Gamma(G+v)}\right)\left(\sum_{q=0}^{\infty}\frac{\Gamma(m+q)E^{q}t^{q}}{q!\Gamma(G+q)}\right). \quad (D.2)$$

The series $b_n$ is a product of two series. We will prove convergence of these two series separately using ratio test. Applying the ratio test for first series in the product,

$$L_{ratio} = \lim_{n\to\infty}\left|\frac{(m+n)Et}{(n+1)(G+n)}\right| = 0 < 1. \quad (D.3)$$

Hence, the first series in the product is convergent. Similarly, convergence of the second series in the product can be proved. From the fact that convergence of two convergent series is convergent, the series $b_n$ is convergent. Since, $a_n < b_n$ and $b_n$ is convergent, from comparison test, $a_n$ is convergent.

## D.2 DERIVATION OF PDF OF SLS UNDER NAKAGAMI FADING IN EQ. (6.17)

We make use of Eq. (6.14) to derive the Eq. (6.17). We first have to derive the CDF of $f_{T_m}(t)$ under $H_m = 1$ given in Eq. (6.7) by integrating it from 0 to $x$. The steps involved are as follows:

$$F_{T_m}(x) = \int_0^x \frac{t^{G-1}e^{-\frac{t}{2}}}{(G-1)!2^G D} {}_1F_1(m_1;G;tE)\,dt. \quad (D.4)$$

DOI: 10.1201/9781003088554-D

The series representation of confluent hypergeometric function given by

$$_1F_1(a,b,z) = \sum_{k=0}^{\infty} \frac{(a)_k}{(b)_k} \frac{z^k}{k!}, \qquad (D.5)$$

where, $(a)_k = \frac{\Gamma(a+k)}{\Gamma(a)}$ represents the Pochhammer symbol, $\Gamma(z)$ represents the Gamma function. Using the series representation of $_1F_1$ in Eq. (D.4) from Eq. (D.5), we get

$$F_{T_m}(x) = \int_0^x \frac{t^{G-1} e^{-\frac{t}{2}}}{(G-1)! 2^G D} \sum_{k=0}^{\infty} \frac{(m_1)_k (tE)^k}{(G)_k k!} dt, \qquad (D.6)$$

After simple mathematical simplifications, we can write

$$F_{T_m}(x) = \sum_{k=0}^{\infty} \frac{(m_1)_k E^k}{(G)_k k! (G-1)! 2^G D} \int_0^x t^{G+k-1} e^{-\frac{t}{2}} dt. \qquad (D.7)$$

Using the change of variable as $\frac{t}{2} = y$, we can write

$$F_{T_m}(x) = \sum_{k=0}^{\infty} \frac{(m_1)_k E^k 2^k}{(G)_k k! (G-1)! D} \int_0^x y^{G+k-1} e^{-y} dy. \qquad (D.8)$$

Note that the integral in the above equation represents $\gamma(G+k, \frac{x}{2})$. Substituting this in above equation we get

$$F_{T_m}(x) = \sum_{k=0}^{\infty} \frac{(m_1)_k E^k 2^k}{(G)_k k! (G-1)! D} \gamma\left(G+k, \frac{x}{2}\right). \qquad (D.9)$$

Substituting Eq. (D.9) after changing dummy variable $x$ to $t$ and pdf from Eq. (6.7) under $H_m = 1$ in Eq. (6.14), we get pdf given in Eq. (6.17) under $H_m = 1$. Similar steps can be applied to get pdf given in Eq. (6.17) under $H_m = 0$. Note that we have used $(G-1)! = \Gamma(G)$ in Eq. (6.17).

## D.3 PROOF FOR CONVERGENCE OF PDF OF SLS UNDER NAKAGAMI FADING IN EQ. (6.17)

The convergence of infinite series in Eq.(6.17) can be proved using the ratio test which is given by

$$L_{ratio} = \lim_{k \to \infty} \left| \frac{C_{k+1}}{C_k} \right|. \qquad (D.10)$$

Here, $C_{k+1}$ and $C_k$ are the $(k+1)^{th}$ and $k^{th}$ terms of the series, respectively. Using Eq.(D.10), one can test the convergence as follows

Appendix for Chapter 6

- If $L_{ratio} < 1$, the series is convergent.
- If $L_{ratio} < 1$, the series is divergent.
- If $L_{ratio} = 1$, the series may be divergent, conditionally convergent, or absolutely convergent.

Now, using the series in Eq.(6.17), $L_{ratio}$ can be obtained as

$$L_{ratio} = \lim_{k \to \infty} \left| \frac{2E(m_1+k)P(G+k+1,\frac{t}{2})}{(k+1)P(G+k,\frac{t}{2})} \right|. \tag{D.11}$$

Using the recurrence formula for normalized incomplete Gamma function, i.e., $P(a+1,x) = P(a,x) - \frac{x^a e^{-x}}{\Gamma(a+1)}$, $L_{ratio}$ can be written as

$$L_{ratio} = \lim_{k \to \infty} \left| \frac{2E(m_1+k)}{(k+1)} \left\{ 1 - \frac{(\frac{t}{2})^{G+k} e^{-\frac{t}{2}}}{(G+k)\gamma(G+k,\frac{t}{2})} \right\} \right|. \tag{D.12}$$

Using the series representation of lower incomplete Gamma function as $\gamma(a,x) \approx \sum_{l=0}^{\infty} \frac{(-1)^l x^{a+l}}{(a+l)l!}$ and taking limit $k \to \infty$, we can show that

$$L_{ratio} = \left| 2E \left\{ 1 - \frac{e^{-\frac{t}{2}}}{\sum_{l=0}^{\infty} \frac{(-1)^l (\frac{t}{2})^l}{l!}} \right\} \right|. \tag{D.13}$$

Using the fact that $\sum_{l=0}^{\infty} \frac{(-1)^l (\frac{t}{2})^l}{l!} = e^{-\frac{t}{2}}$, we can obtain $L_{ratio}$ as

$$L_{ratio} = \lim_{k \to \infty} \left| \frac{C_{k+1}}{C_k} \right| = 0. \tag{D.14}$$

Since, $L_{ratio}$ in Eq.(D.14) is $<1$, the series given in Eq.(6.17) is convergent.

## D.4 DERIVATION OF PDF IN EQ. (6.13)

The pdf of sum of two independent random variables $X$ and $Y$. i.e., $Z = X + Y$, can be obtained by convolving the marginal pdfs of $X$ and $Y$, which can be obtained as

$$f_Z(z) = \int_{-\infty}^{\infty} f_X(x) f_Y(z-x) dx. \tag{D.15}$$

First we derive the pdf of decision statistic under SLC diversity for $H_m = 1$. After substituting pdf from Eq. (6.11) under $H_m = 1$ into Eq. (D.15), we obtain

$$f_Z(z) = a^2 e^{-\alpha z} \int_0^z x^q (z-x)^q dx. \tag{D.16}$$

After performing the convolution, we obtain the pdf of decision statistic under SLC diversity with two diversity branches. It is given as

$$f_Z(z) = \frac{a^2 \Gamma(1+q)^2 z^{2q+1} e^{-\alpha z}}{\Gamma(2q+2)}. \tag{D.17}$$

The pdf of decision statistic under SLC diversity with three diversity branches can be obtained by convolving pdf in Eq. (D.17) with Eq. (6.11) under $H_m = 1$. Following the similar procedure of convolution as done for two diversity case, the pdf for the case of three diversity branch can be obtained as

$$f_z(z) = \frac{a^3 \Gamma(1+q)^3 z^{3q+2} e^{-az}}{\Gamma(3q+3)}. \tag{D.18}$$

Looking at the pattern in Eq. (D.17) and Eq. (D.18), we can write the pdf of decision statistic under SLC diversity with $P$ diversity branches as

$$f_{T_m}(t) = \frac{a^P \Gamma(1+q)^P t^{Pq+P-1} e^{-at}}{\Gamma(Pq+P)}. \tag{D.19}$$

The pdf of decision statistic under $H_m = 0$ follows chi-squared distribution. If $Y = X_1 + X_2 + \cdots + X_P$, where $X_1, X_2, \ldots, X_P$ are following chi-squared distribution with $N$ degrees of freedom then $Y$ follows chi-squared distribution with $PN$ degrees of freedom. Hence, the pdf of decision statistic using SLC diversity under $H_m = 0$ with $P$ diversity branches is given as

$$f_{T_m}(t) = \chi^2_{PN} = \frac{t^{P \cdot G - 1} e^{-\frac{t}{2}}}{2^{P \cdot G} \Gamma(P \cdot G)}. \tag{D.20}$$

## D.5 DERIVATION OF EQ. (6.34)

The series representation of upper incomplete Gamma function can be written as

$$\Gamma(n, x) = (n-1)! e^{-x} \sum_{k=0}^{n-1} \frac{x^k}{k!}. \tag{D.21}$$

Using series representation of upper incomplete Gamma function from Eq. (D.21) in Eq. (6.33) and integrating it from 0 to $\tau$ we get

$$PS_{00}^{Nak}(\tau) = \int_0^\tau \frac{t^{PG-1} e^{-\frac{t}{2}}}{2^{PG-1} \Gamma(PG)^2} (PG-1)! e^{-\frac{t}{2}} \sum_{k=0}^{PG-1} \frac{\left(\frac{t}{2}\right)^2}{k!} dt. \tag{D.22}$$

Rearranging the Eq. (D.22), we can write

$$PS_{00}^{Nak}(\tau) = \sum_{k=0}^{PG-1} \frac{(PG-1)!}{2^{PG-1} 2^k k! \Gamma(PG)^2} \int_0^\tau t^{PG+k-1} e^{-t} dt. \tag{D.23}$$

Note that the inner integral represents $\gamma(PG+k, \tau)$ and $(PG-1)! = \Gamma(PG)$. Using these in Eq. (D.23), we get $PS_{00}^{Nak}(\tau)$ as

$$PS_{00}^{Nak}(\tau) = \sum_{k=0}^{P \cdot G - 1} \frac{\gamma(P \cdot G + k, \tau)}{2^{PG+k-1} k! \Gamma(P \cdot G)}. \tag{D.24}$$

# Appendix for Chapter 6

## D.6 DERIVATION OF PDF IN EQ. (6.36)

Using $L = 2$ in Eq. (6.32), we can write the expression for $f_{T_{min}}(t)$ as

$$f_{T_{min}}(t) = 2f_T(t)(1 - F_T(t)). \qquad (D.25)$$

From Eq. (D.25), we can see that one needs to derive cdf $F_T(t)$ in order to derive $f_{T_{min}}(t)$ which can be obtained by integrating $f_{T_m}(t)$ under $H_m = 1$ in Eq. (6.13) from 0 to $t$. We can derive $F_T(t)$ as follows

$$F_T(t) = \int_0^t \frac{a^P \Gamma(1+q)^P x^{Pq+P-1} e^{-\alpha x}}{\Gamma(Pq+P)} dt. \qquad (D.26)$$

Note that we have used dummy variable $x$ in place of $t$ in Eq. (6.13). Rearranging the above equation we can write

$$F_T(t) = \frac{a^P \Gamma(1+q)^P}{\Gamma(Pq+P)} \int_0^t x^{Pq+P-1} e^{-\alpha x} dt. \qquad (D.27)$$

Taking the change of variable as $y = \alpha x$, we get

$$F_T(t) = \frac{a^P \Gamma(1+q)^P}{\Gamma(Pq+P)\alpha^{Pq+P}} \int_0^{\alpha t} y^{Pq+P-1} e^{-y} dt. \qquad (D.28)$$

The integral term in above equation can be written as $\gamma(Pq+P, \alpha t)$. Using tis we get $F_T(t)$ as

$$F_T(t) = \frac{a^P \Gamma(1+q)^P \gamma(Pq+P, \alpha t)}{\Gamma(Pq+P)\alpha^{Pq+P}}. \qquad (D.29)$$

Substituting Eq. (D.29) and pdf from Eq. (6.13) under $H_m = 1$ in Eq. (D.25), we get the expression given in Eq. (6.36).

## D.7 THEORETICAL ANALYSIS OF R-SLC FOR $L = 3$

Consider the case when there are three channels in the partial band, i.e., $L = 3$ with $S_d = L_d = 1$, $I_d = 0$. Based on the number of ON channels, there are four cases, i.e., all channels OFF, only one channel ON, two channels ON and all the channels ON. First, let us consider that all the channels are OFF. In this case $PS_{000}^{Nak}(\tau)$ can be obtained as

$$PS(\tau|H_1, H_2, H_3 = 0) = PS_{000}^{Nak}(\tau) = Pr\{T_{min} < \tau\}. \qquad (D.30)$$

The pdf of $T_{min}$, i.e., $f_{T_{min}}(t)$, for this case can be obtained by using pdf under $H_m = 0$ from Eq. (6.13) in Eq. (6.32) and using $L = 3$. Once pdf of $T_{min}$ is obtained, $PS_{000}^{Nak}(\tau)$

can be obtained by integrating $f_{T_{min}}(t)$ from 0 to $\tau$. Since all the channels are ON, there is no chance of interference and hence the probability of interference opportunity is zero, i.e., $PI_{000}^{Nak}(\tau) = 0$.

We now consider the case when all the channels are ON. In this case $PI_{111}^{Nak}(\tau)$ can be obtained as

$$PI(\tau|H_1, H_2, H_3 = 1) = PI_{111}^{Nak}(\tau) = Pr\{T_{min} < \tau\}. \quad (D.31)$$

The $f_{T_{min}}(t)$ in this case can be obtained in a similar way corresponding to OFF case by using pdf under $H_m = 1$ from Eq. (6.13). $PS_{111}^{Nak}(\tau)$ is zero since all the channels are ON.

Next we consider the case of only one channel being ON. Assuming channels 1 and 2 as OFF and channel 3 as ON, $PS_{001}^{Nak}(\tau)$ can be obtained as

$$PS(\tau|H_1 = H_2 = 0, H_3 = 1) = PS_{001}^{Nak}(\tau),$$
$$= Pr\{T_1 < (T_2, T_3), T_1 < \tau\} + Pr\{T_2 < (T_1, T_3), T_2 < \tau\},$$
$$= \int_0^\tau \left[ f_{T_1}(t_1) \left( \int_{t_1}^\infty f_{T_2}(t_2)\, dt_2 \right) \left( \int_{t_1}^\infty f_{T_3}(t_3)\, dt_3 \right) \right] dt_1$$
$$+ \int_0^\tau \left[ f_{T_2}(t_2) \left( \int_{t_2}^\infty f_{T_1}(t_1)\, dt_1 \right) \left( \int_{t_2}^\infty f_{T_3}(t_3)\, dt_3 \right) \right] dt_2. \quad (D.32)$$

Since channels 1 and 2 are OFF, $f_{T_1}(t_1) = f_{T_2}(t_2) = f_T(t)$. Using this, $PS_{001}^{Nak}(\tau)$ reduces to

$$PS_{001}^{Nak}(\tau) = 2 \int_0^\tau f_T(t)\, \bar{F}_T(t)\, \bar{F}_{T_3}(t)\, dt, \quad (D.33)$$

where, $\bar{F}_T(t)$ and $\bar{F}_{T_3}(t)$ are the ccdfs for $f_T(t)$ and $f_{T_3}(t_3)$, respectively. The PI in this case can be obtained as

$$PI(\tau|H_1, H_2 = 0, H_3 = 1) = PI_{001}^{Nak}(\tau),$$
$$= Pr\{T_3 < (T_1, T_2), T_3 < \tau\},$$
$$= \int_0^\tau f_{T_3}(t)\, (\bar{F}_T(t))^2\, dt. \quad (D.34)$$

Finally, we consider only one channel being OFF. Let us consider channel 1 as OFF and channels 2 and 3 as ON. The PS is then obtained as

$$PS(\tau|H_1 = 0, H_2 = H_3 = 1) = PS_{011}^{Nak}(\tau)$$
$$= Pr\{T_1 < (T_2, T_3), T_1 < \tau\},$$
$$= \int_0^\tau f_{T_1}(t)\, (\bar{F}_T(t))^2\, dt. \quad (D.35)$$

Since channels 2 and 3 are ON, $f_{T_2}(t_2) = f_{T_3}(t_3) = f_T(t)$. Following a similar procedure, the PI in this case can be obtained as

$$PI_{011}^{Nak}(\tau) = 2 \int_0^\tau f_T(t)\bar{F}_T(t)\bar{F}_{T_1}(t)\, dt. \quad (D.36)$$

# Appendix for Chapter 6

Once PS and PI are known for all four cases, $P_{ISO}$ and $P_{EIO}$ can be obtained as

$$P_{ISO}^{Nak}(\tau) = 1 - \sum_{i=0}^{1}\sum_{j=0}^{1}\sum_{k=0}^{1}(1-p)^{L-i-j-k}p^{i+j+k}PS_{ijk}^{Nak}(\tau), \qquad (D.37)$$

$$P_{EIO}^{Nak}(\tau) = \sum_{i=0}^{1}\sum_{j=0}^{1}\sum_{k=0}^{1}(1-p)^{L-i-j-k}p^{i+j+k}PI_{ijk}^{Nak}(\tau), \qquad (D.38)$$

respectively.

In a similar way, the analysis can be done for any parameter setting. Note that when we consider $L$ channels in the partial band, we need to find $PS$ and $PI$ considering $\binom{L}{1}+1$ cases. For example, for $L=4$, we need to consider five cases, i.e., all channels OFF, any one channel ON, any two channels ON, any three channels ON and all four channels ON.

# E Some Special Functions

In this APPENDIX, the special functions used in this book are discussed in brief. Although the discussion given here will be sufficient to understand the derivation given in this book, readers are advised not to consider this as a detailed discussion on these topics. Readers are advised to refer books dedicated to a special function for more details.

## E.1 GAMMA FUNCTION

The integral form of Gamma function is represented as

$$\Gamma(z) = \int_0^\infty t^{z-1} e^{-t} dt \quad (\mathscr{R}(z) > 0). \tag{E.1}$$

The relation of Gamma function with factorial is given by the following relation

$$\Gamma(z+1) = z\Gamma(z) = z! = z(z-1)!. \tag{E.2}$$

## E.2 LOWER INCOMPLETE GAMMA FUNCTION

The regularized lower incomplete Gamma function is defined as

$$P(a,x) = \frac{1}{\Gamma(a)} \int_0^x t^{a-1} e^{-t} dt, \quad (\mathscr{R}(a) > 0). \tag{E.3}$$

The lower incomplete Gamma function has the following relation with regularized lower incomplete Gamma function

$$\gamma(a,x) = P(a,x)\Gamma(a) = \int_0^x t^{a-1} e^{-t} dt, \quad (\mathscr{R}.(a) > 0) \tag{E.4}$$

The recurrence formulas for lower incomplete gamma functions are as follows

$$P(a+1,x) = P(a,x) - \frac{x^a e^{-x}}{\Gamma(a+1)}. \tag{E.5}$$

$$\gamma(a+1,x) = a\gamma(a,x) - x^a e^{-x}. \tag{E.6}$$

For $a$ as an integer $n$, the series expansion of $\gamma(a,x)$ is given as

$$\gamma(n,x) = (n-1)! \left(1 - e^{-x} \sum_{k=0}^{n-1} \frac{x^k}{k!}\right). \tag{E.7}$$

## E.3 UPPER INCOMPLETE GAMMA FUNCTION

The regularized upper incomplete Gamma function is defined as

$$Q(a,x) = \frac{1}{\Gamma(a)} \int_x^\infty t^{a-1} e^{-t} dt, \quad (\mathscr{R}(a) > 0). \tag{E.8}$$

The regularized lower and upper incomplete Gamma functions satisfy the following relation

$$P(a,x) = Q(a,x) = 1. \tag{E.9}$$

The upper incomplete Gamma function has the following relation with regularized upper incomplete Gamma function

$$\Gamma(a,x) = Q(a,x)\Gamma(a) = \int_x^\infty t^{a-1} e^{-t} dt, \quad (\mathscr{R}(a) > 0). \tag{E.10}$$

The upper and lower incomplete Gamma functions are related by the formula

$$\Gamma(a,x) = \Gamma(a) - \gamma(a,x). \tag{E.11}$$

For $a$ an integer $n$, the series expansion of $\Gamma(a,x)$ is given as

$$\Gamma(n,x) = (n-1)! \left( e^{-x} \sum_{k=0}^{n-1} \frac{x^k}{k!} \right). \tag{E.12}$$

## E.4 GENERALIZED MARCUM Q-FUNCTION

The generalized Marcum Q-function is defined as

$$Q_M(\alpha, \beta) = \frac{1}{\alpha^{M-1}} \int_\beta^\alpha x^M e^{-\frac{(x^2+\alpha^2)}{2}} I_{M-1}(\alpha x) dx, \tag{E.13}$$

where, $I_n(x)$ is a modified Bessel function of the first kind. The generalized Marcum Q-function has the following series form

$$Q_M(\alpha, \beta) = e^{-\frac{(\alpha^2+\beta^2)}{2}} \sum_{k=1-M}^\infty \left( \frac{\alpha}{\beta} \right)^k I_k(\alpha\beta). \tag{E.14}$$

## E.5 BESSEL FUNCTION OF THE FIRST KIND

The Bessel functions of the first kind $J_n(x)$ are defined as the solutions of the Bessel differential equation

$$x^2 \frac{d^2 y}{dx^2} + x \frac{dy}{dx} + (x^2 - n^2) y = 0, \tag{E.15}$$

which are nonsingular at the origin. The Bessel function $J_n(x)$ can be defined using the contour integral as

$$J_n(x) = \frac{1}{2\pi j} \oint e^{\left(\frac{x}{2}\right)\left(\frac{t-1}{t}\right)} t^{-n-1} dt, \qquad (E.16)$$

where, the contour encloses the origin and is traversed in a counterclockwise direction [13]. The series form of Bessel function is given as

$$J_m(x) = \sum_{l=0}^{\infty} \frac{(-1)^l}{2^{2l+m} l! (m+l)!} x^{2l+m}, \qquad (E.17)$$

where, the factorials can be generalized to gamma functions for non-integer $m$.

## E.6 MODIFIED BESSEL FUNCTION OF THE FIRST KIND

The modified Bessel function of the first kind $I_n(x)$ can be defined by the contour integral as

$$I_n(x) = \frac{1}{2\pi i} \oint e^{\left(\frac{x}{2}\right)\left(\frac{t+1}{t}\right)} t^{-n-1} dt, \qquad (E.18)$$

where, the contour encloses the origin and is traversed in a counterclockwise direction [13]. The modified Bessel function of first kind is related with the Bessel function of first kind by the following formula

$$I_n(x) = i^{-n} J_n(ix) = e^{-\frac{n\pi i}{2}} J_n\left(xe^{\frac{i\pi}{2}}\right). \qquad (E.19)$$

The series representation of modified Bessel function of first kind for real number $v$ can be given as

$$I_v(x) = \left(\frac{1}{2}x\right)^v \sum_{k=0}^{\infty} \frac{\left(\frac{1}{4}x^2\right)^k}{k! \Gamma(v+k+1)}, \qquad (E.20)$$

where, $\Gamma(x)$ is the Gamma function.

## E.7 CONFLUENT HYPERGEOMETRIC FUNCTION

The confluent hypergeometric function has a hypergeometric series given by [79]

$$_1F_1(a,b,x) = 1 + \frac{a}{b}x + \frac{a(a+1)}{b(b+1)}\frac{x^2}{2!} + \cdots = \sum_{k=0}^{\infty} \frac{(a)_k}{(b)_k}\frac{x^k}{k!}, \qquad (E.21)$$

where, $(a)_k$ and $(b)_k$ represent Pochhammer symbols defined as below

$$(x)_n = \frac{\Gamma(x+n)}{\Gamma(n)} = x(x+1)\cdots(x+n-1). \qquad (E.22)$$

The confluent hypergeometric function also has the following integral form

$$_1F_1(a,b,x) = \frac{\Gamma(b)}{\Gamma(b-a)\Gamma(a)} \int_0^1 e^{xt} t^{a-1}(1-t)^{b-a-1} dt. \qquad (E.23)$$

## E.8 CONFLUENT HYPERGEOMETRIC FUNCTION OF THE SECOND KIND

The confluent hypergeometric function of second Kind has the following integral form

$$U(a,b,x) = \frac{1}{\Gamma(a)} \int_0^\infty e^{-xt} t^{a-1} (1+t)^{b-a-1} dt, \qquad (E.24)$$

for $\mathscr{R}(a), \mathscr{R}(b) > 0$ [13].

## E.9 UNIT STEP FUNCTION

The continuous time unit step function is denoted by $1(t)$ or $u(t)$ which is defined as follows

$$u(t) = \begin{cases} 0, & t < 0 \\ 1, & t > 0. \end{cases} \qquad (E.25)$$

The unit step function is discontinuous at $t = 0$.

## E.10 Q-FUNCTION

The Q-function is defined as

$$Q(x) = \frac{1}{\sqrt{2\pi}} \int_x^\infty e^{-\frac{y^2}{2}} dy. \qquad (E.26)$$

The Q-function is a monotonically decreasing function. Some of the values it takes are as follows

$$Q(-\infty) = 1; \quad Q(0) = \frac{1}{2}; \quad Q(\infty) = 0. \qquad (E.27)$$

The Q-function satisfies the following property

$$Q(x) = 1 - Q(-x) = 1 - \Phi(x), \qquad (E.28)$$

where, $\Phi(x)$ is the cumulative distribution function of the standard normal Gaussian distribution $\mathscr{N}(0,1)$.

If we have a normal random variable $X \sim \mathscr{N}(\mu, \sigma^2)$, the probability of $X > x$ can be obtained as

$$Pr(X > x) = Q\left(\frac{x-\mu}{\sigma}\right). \qquad (E.29)$$

# Some Special Functions

## E.11 ERROR FUNCTION

The integral form of the error function is given as

$$erf(x) = \frac{2}{\sqrt{\pi}} \int_0^x e^{-y^2} dy. \tag{E.30}$$

The error function has the following values

$$erf(0) = 0 \text{ and } erf(\infty) = 1. \tag{E.31}$$

The error function is an odd function, i.e.,

$$erf(x) = -erf(x). \tag{E.32}$$

The Maclaurin series form of $erf(x)$ can be given as

$$erf(x) = \frac{2}{\sqrt{\pi}} \sum_{n=0}^{\infty} \frac{(-1)^n x^{2n+1}}{n!(2n+1)}. \tag{E.33}$$

The integral form of complementary error function is given by

$$erfc(x) = \frac{2}{\sqrt{\pi}} \int_x^{\infty} e^{-y^2} dy. \tag{E.34}$$

The relationship between error function (erf) and complementary error function (erfc) is given by

$$erf(x) = 1 - erfc(x). \tag{E.35}$$

The relation between Q-function and complementary error function is given by the following relation

$$Q(x) = \frac{1}{2} erfc\left(\frac{x}{\sqrt{2}}\right). \tag{E.36}$$

The error function and the confluent hypergeometric function of first kind are related by the following formula

$$erf(x) = \frac{2x}{\sqrt{\pi}} {}_1F_1\left(\frac{1}{2}; \frac{3}{2}; -x^2\right) = \frac{2xe^{-x^2}}{\sqrt{\pi}} {}_1F_1\left(1; \frac{3}{2}; x^2\right). \tag{E.37}$$

For $x > 0$, the lower incomplete Gamma function is related to error function as

$$erf(x) = \pi^{-\frac{1}{2}} \gamma\left(\frac{1}{2}, x^2\right). \tag{E.38}$$

## E.12 POLYLOGARITHM

The polylogarithm $Li_n(x)$ is defined by the power series in $x$ as

$$Li_n(x) = \sum_{k=1}^{\infty} \frac{x^k}{k^n} = x + \frac{x^2}{2^n} + \frac{x^3}{3^n} + \cdots. \tag{E.39}$$

# Bibliography

1. "UK spectrum map", static.ofcom.org.uk, 2020. [Online]. Available: http://static.ofcom.org.uk/static/spectrum/map.html. [Accessed: 25 Feb 2020].
2. "FCC Allocation History File", Federal Communications Commission Office of Engineering and Technology Policy and Rules Division. May 2019.
3. A. Abdi and M. Kaveh. K distribution: an appropriate substitute for rayleigh-lognormal distribution in fading-shadowing wireless channels. *Electronics Letters*, 34(9):851–852, Apr 1998.
4. M. Abramovitz and I. A. Stigan, eds. *Handbook of Mathematical Functions*. New York. NY: Dover, 1970.
5. I. F. Akyildiz, W.Y. Lee, M. C. Vuran, and S. Mohanty. Next generation/dynamic spectrum access/cognitive radio wireless networks: A survey. *Elsevier Computer Networks*, 50(13):2127–2159, 2006.
6. Ian F. Akyildiz, Won-Yeol Lee, Mehmet C. Vuran, and Shantidev Mohanty. Next generation/dynamic spectrum access/cognitive radio wireless networks: A survey. *Computer Networks*, 50(13):2127–2159, 2006.
7. Ian F. Akyildiz, Brandon F. Lo, and Ravikumar Balakrishnan. Cooperative spectrum sensing in cognitive radio networks: A survey. *Physical Communication*, 4(1):40 – 62, 2011.
8. H. Al-Hmood and H. S. Al-Raweshidy. Unified modeling of composite $\kappa-\mu$/gamma, $\eta-\mu$/gamma, and $\alpha-\mu$/gamma fading channels using a mixture gamma distribution with applications to energy detection. *IEEE Antennas and Wireless Propagation Letters*, 16:104–108, 2017.
9. H. Al-Hmood and H. S. Al-Raweshidy. On the effective rate and energy detection based spectrum sensing over $\alpha-\eta-\kappa-\mu$ fading channels. *IEEE Transactions on Vehicular Technology*, 69(8):9112–9116, 2020.
10. O. Alhussein, A. A. Hammadi, P. C. Sofotasios, S. Muhaidat, J. Liang, M. Al-Qutayri, and G. K. Karagiannidis. Performance analysis of energy detection over mixture gamma based fading channels with diversity reception. In *2015 IEEE 11th International Conference on Wireless and Mobile Computing, Networking and Communications (WiMob)*, pages 399–405, Oct 2015.
11. A. Ali and W. Hamouda. Advances on spectrum sensing for cognitive radio networks: Theory and applications. *IEEE Communications Surveys Tutorials*, 19(2):1277–1304, Secondquarter 2017.
12. T. An, I. Song, S. Lee, and H. K. Min. Detection of signals with observations in multiple subbands: A scheme of wideband spectrum sensing for cognitive radio with multiple antennas. *IEEE Transactions on Wireless Communications*, 13(12):6968–6981, Dec 2014.
13. George B. Arfken and Hans J. Weber. *Mathematical Methods for Physicists*. Elsevier Academic Press, San Diego, 6th edition, 2005.

14. D.D. Ariananda and G. Leus. Compressive wideband power spectrum estimation. *IEEE Transactions on Signal Processing,* 60(9):4775–4789, Sept 2012.
15. D.D. Ariananda, G. Leus, and Zhi Tian. Multi-coset sampling for power spectrum blind sensing. In *17th International Conference on Digital Signal Processing (DSP),* pages 1–8, July 2011.
16. A. Assra, J. Yang, and B. Champagne. An EM approach for cooperative spectrum sensing in multiantenna CR networks. *IEEE Transactions on Vehicular Technology,* 65(3):1229–1243, March 2016.
17. S. Atapattu, C. Tellambura, and H. Jiang. Spectrum sensing via energy detector in low SNR. In *2011 IEEE International Conference on Communications (ICC),* pages 1–5, June 2011.
18. S. Atapattu, C. Tellambura, and Hai Jiang. Energy detection of primary signals over $\eta$-$\mu$ fading channels. In *International Conference on Industrial and Information Systems (ICIIS),* pages 118–122, Dec 2009.
19. S. Atapattu, C. Tellambura, and Hai Jiang. Performance of an energy detector over channels with both multipath fading and shadowing. *IEEE Transactions on Wireless Communications,* 9(12):3662–3670, December 2010.
20. E. Axell, G. Leus, E. G. Larsson, and H. V. Poor. Spectrum sensing for cognitive radio: State-of-the-art and recent advances. *IEEE Signal Processing Magazine,* 29(3):101–116, 2012.
21. V. Badrinarayanan, A. Kendall, and R. Cipolla. Segnet: A deep convolutional encoder-decoder architecture for image segmentation. *IEEE Transactions on Pattern Analysis and Machine Intelligence,* 39(12):2481–2495, 2017.
22. A. Bagheri, P. C. Sofotasios, T. A. Tsiftsis, K. Ho-Van, M. I. Loupis, S. Freear, and M. Valkama. Energy detection based spectrum sensing over enriched multipath fading channels. In *2016 IEEE Wireless Communications and Networking Conference,* pages 1–6, April 2016.
23. T. W. Ban, W. Choi, B. C. Jung, and D. K. Sung. Multi-user diversity in a spectrum sharing system. *IEEE Transactions on Wireless Communications,* 8(1):102–106, 2009.
24. V. R. Sharma Banjade, C. Tellambura, and H. Jiang. Performance of p-norm detector in AWGN, fading, and diversity reception. *IEEE Transactions on Vehicular Technology,* 63(7):3209–3222, Sept 2014.
25. V. R. Sharma Banjade, C. Tellambura, and H. Jiang. Approximations for performance of energy detector and $p$-norm detector. *IEEE Communications Letters,* 19(10):1678–1681, Oct 2015.
26. V. R. Sharma Banjade, C. Tellambura, and H. Jiang. Asymptotic performance of energy detector in fading and diversity reception. *IEEE Transactions on Communications,* 63(6):2031–2043, June 2015.
27. Vesh Raj Sharma Banjade, Chintha Tellambura, and Hai Jiang. New asymptotics for performance of energy detector. In *IEEE Global Communications Conference (GLOBECOM),* pages 4020–4024, 2014.
28. D. Bhargavi and C. R. Murthy. Performance comparison of energy, matched-filter and cyclostationarity-based spectrum sensing. In *2010 IEEE 11th*

*International Workshop on Signal Processing Advances in Wireless Communications (SPAWC)*, pages 1–5, 2010.
29. K. Bouallegue, I. Dayoub, M. Gharbi, and K. Hassan. Blind spectrum sensing using extreme eigenvalues for cognitive radio networks. *IEEE Communications Letters*, 22(7):1386–1389, 2018.
30. A. A. A. Boulogeorgos, N. D. Chatzidiamantis, and G. K. Karagiannidis. Energy detection spectrum sensing under rf imperfections. *IEEE Transactions on Communications*, 64(7):2754–2766, July 2016.
31. D. Cabric, S. M. Mishra, and R. W. Brodersen. Implementation issues in spectrum sensing for cognitive radios. In *Conference Record of the Thirty-Eighth Asilomar Conference on Signals, Systems and Computers, 2004*, volume 1, pages 772–776, Nov 2004.
32. K. Captain and M. Joshi. Exploiting space and antenna diversity for wideband spectrum sensing. In *2021 International Conference on COMmunication Systems NETworkS (COMSNETS)*, pages 179–183, 2021.
33. K. M. Captain and M. V. Joshi. Performance of wideband spectrum sensing under fading channel. In *2015 Seventh International Workshop on Signal Design and its Applications in Communications (IWSDA)*, pages 170–174, Sept 2015.
34. K. M. Captain and M. V. Joshi. Energy detection based spectrum sensing over $\eta$-$\lambda$-$\mu$ fading channel. In *8th International Conference on Communication Systems and Networks (COMSNETS)*, pages 1–6, Jan 2016.
35. K. M. Captain and M. V. Joshi. Square law selection diversity for wideband spectrum sensing under fading. In *2016 IEEE 84th Vehicular Technology Conference (VTC-Fall)*, pages 1–6, Sept 2016.
36. K. M. Captain and M. V. Joshi. Performance improvement in wideband spectrum sensing under fading: Use of diversity. *IEEE Transactions on Vehicular Technology*, 66(9):8152–8162, Sept 2017.
37. K. M. Captain and M. V. Joshi. SNR wall for cooperative spectrum sensing using generalized energy detector. In *Accepted for Publication at 10th International Conference on Communication Systems and Networks (COMSNETS)*, Jan 2018.
38. K. M. Captain and M. V. Joshi. A PBNS based detection algorithm for cooperative wideband spectrum sensing using hard combining. In *2019 IEEE 89th Vehicular Technology Conference (VTC2019-Spring)*, pages 1–5, April 2019.
39. T. Chan, K. Jia, S. Gao, J. Lu, Z. Zeng, and Y. Ma. PCANet: A simple deep learning baseline for image classification? *IEEE Transactions on Image Processing*, 24(12):5017–5032, 2015.
40. G. Chandrasekaran and S. Kalyani. Performance analysis of cooperative spectrum sensing over $\kappa-\mu$ shadowed fading. *IEEE Wireless Communications Letters*, 4(5):553–556, 2015.
41. E. Chatziantoniou, B. Allen, V. Velisavljevic, P. Karadimas, and J. Coon. Energy detection based spectrum sensing over two-wave with diffuse power fading channels. *IEEE Transactions on Vehicular Technology*, 66(1):868–874,

2017.

42. R. B. Chaurasiya and R. Shrestha. Hardware-efficient and fast sensing-time maximum-minimum-eigenvalue-based spectrum sensor for cognitive radio network. *IEEE Transactions on Circuits and Systems I: Regular Papers*, 66(11):4448–4461, 2019.

43. R. B. Chaurasiya and R. Shrestha. Fast sensing-time and hardware-efficient eigenvalue-based blind spectrum sensors for cognitive radio network. *IEEE Transactions on Circuits and Systems I: Regular Papers*, 67(4):1296–1308, 2020.

44. D. Chen, J. Li, and J. Ma. Cooperative spectrum sensing under noise uncertainty in cognitive radio. In *2008 4th International Conference on Wireless Communications, Networking and Mobile Computing*, pages 1–4, Oct 2008.

45. H. S. Chen, W. Gao, and D. G. Daut. Spectrum sensing using cyclostationary properties and application to IEEE 802.22 WRAN. In *IEEE Global Telecommunications Conference (GLOBECOM)*, pages 3133–3138, Nov 2007.

46. X. Chen, W. Xu, Z. He, and X. Tao. Spectral correlation-based multi-antenna spectrum sensing technique. In *2008 IEEE Wireless Communications and Networking Conference*, pages 735–740, March 2008.

47. Y. Chen. Improved energy detector for random signals in Gaussian noise. *IEEE Transactions on Wireless Communications*, 9(2):558–563, February 2010.

48. Y. Chen, Z. Lin, X. Zhao, G. Wang, and Y. Gu. Deep learning-based classification of hyperspectral data. *IEEE Journal of Selected Topics in Applied Earth Observations and Remote Sensing*, 7(6):2094–2107, 2014.

49. W. Chin. On the noise uncertainty for the energy detection of OFDM signals. *IEEE Transactions on Vehicular Technology*, 68(8):7593–7602, 2019.

50. Y. Chu and S. Liu. Hard decision fusion based cooperative spectrum sensing over Nakagami-m fading channels. In *2012 8th International Conference on Wireless Communications, Networking and Mobile Computing*, pages 1–4, Sept 2012.

51. D. Cohen and Y.C. Eldar. Sub-Nyquist sampling for power spectrum sensing in cognitive radios: A unified approach. *IEEE Transactions on Signal Processing*, 62(15):3897–3910, Aug 2014.

52. Carlos Cordeiro, Kiran Challapali, Dagnachew Birru, and Sai Shankar N. IEEE 802.22: An introduction to the first wireless standard based on cognitive radios. *Journal of Communications*, 1:38–47, April 2006.

53. A. V. Dandawate and G. B. Giannakis. Statistical tests for presence of cyclostationarity. *IEEE Transactions on Signal Processing*, 42(9):2355–2369, Sep 1994.

54. K. Davaslioglu and Y. E. Sagduyu. Generative adversarial learning for spectrum sensing. In *2018 IEEE International Conference on Communications (ICC)*, pages 1–6, 2018.

55. M. Derakhtian, F. Izedi, A. Sheikhi, and M. Neinavaie. Cooperative wideband spectrum sensing for cognitive radio networks in fading channels. *IET Signal Processing*, 6(3):227–238, May 2012.

# BIBLIOGRAPHY

56. M. Derakhtian, F. Izedi, A. Sheikhi, and M. Neinavaie. Cooperative wideband spectrum sensing for cognitive radio networks in fading channels. *IET Signal Processing*, 6(3):227–238, May 2012.
57. Natasha Devroye and Vahid Tarokh. *Fundamental Limits of Cognitive Radio Networks*, pages 327–351. Springer Netherlands, Dordrecht, 2007.
58. Y. Dhungana and C. Tellambura. New simple approximations for error probability and outage in fading. *IEEE Communications Letters*, 16(11):1760–1763, November 2012.
59. F. F. Digham, M. S. Alouini, and M. K. Simon. On the energy detection of unknown signals over fading channels. In *IEEE International Conference on Communications*, volume 5, pages 3575–3579, May 2003.
60. F.F. Digham, M.-S. Alouini, and Marvin K. Simon. On the energy detection of unknown signals over fading channels. *IEEE Transactions on Communications*, 55(1):21–24, Jan 2007.
61. J. Donahue, L. A. Hendricks, S. Guadarrama, M. Rohrbach, S. Venugopalan, T. Darrell, and K. Saenko. Long-term recurrent convolutional networks for visual recognition and description. In *2015 IEEE Conference on Computer Vision and Pattern Recognition (CVPR)*, pages 2625–2634, 2015.
62. S. E. El-Khamy, M. S. El-Mahallawy, and E. N. S. Youssef. Improved wideband spectrum sensing techniques using wavelet-based edge detection for cognitive radio. In *2013 International Conference on Computing, Networking and Communications (ICNC)*, pages 418–423, Jan 2013.
63. B. Farhang-Boroujeny. Filter bank spectrum sensing for cognitive radios. *IEEE Transactions on Signal Processing*, 56(5):1801–1811, May 2008.
64. FCC. Spectrum policy task force report (ET Docket no. 02-135). Nov 2002.
65. FCC. Spectrum policy task force report (ET Docket no. 02-155). Nov 2002.
66. FCC. Notice of proposed rule making and order (ET Docket no. 03-322). Dec 2003.
67. Y. Feng and X. Wang. Adaptive multiband spectrum sensing. *IEEE Wireless Communications Letters*, 1(2):121–124, April 2012.
68. G. Ganesan and Y. Li. Cooperative spectrum sensing in cognitive radio, part i: Two user networks. *IEEE Transactions on Wireless Communications*, 6(6):2204–2213, June 2007.
69. G. Ganesan and Y. Li. Cooperative spectrum sensing in cognitive radio, part ii: Multiuser networks. *IEEE Transactions on Wireless Communications*, 6(6):2214–2222, June 2007.
70. J. Gao, X. Yi, C. Zhong, X. Chen, and Z. Zhang. Deep learning for spectrum sensing. *IEEE Wireless Communications Letters*, 8(6):1727–1730, 2019.
71. Y. Gao, Y. Chen, and Y. Ma. Sparse-bayesian-learning-based wideband spectrum sensing with simplified modulated wideband converter. *IEEE Access*, 6:6058–6070, 2018.
72. W. A. Gardner. Signal interception: a unifying theoretical framework for feature detection. *IEEE Transactions on Communications*, 36(8):897–906, Aug 1988.

73. W. A. Gardner. Exploitation of spectral redundancy in cyclostationary signals. *IEEE Signal Processing Magazine*, 8(2):14–36, April 1991.
74. A. Ghasemi and E. S. Sousa. Collaborative spectrum sensing for opportunistic access in fading environments. In *First IEEE International Symposium on New Frontiers in Dynamic Spectrum Access Networks, 2005. DySPAN 2005*, pages 131–136, Nov 2005.
75. A. Ghasemi and E.S. Sousa. Impact of user collaboration on the performance of sensing-based opportunistic spectrum access. In *Vehicular Technology Conference, 2006. VTC-2006 Fall. 2006 IEEE 64th*, pages 1–6, Sept 2006.
76. M. Ghaznavi and A. Jamshidi. Efficient method for reducing the average control bits in a distributed cooperative sensing in cognitive radio system. *IET Communications*, 7(9):867–874, June 2013.
77. A. Ghosh and W. Hamouda. On the performance of interference-aware cognitive Ad-Hoc networks. *IEEE Communications Letters*, 17(10):1952–1955, 2013.
78. A. Goldsmith, S. A. Jafar, I. Maric, and S. Srinivasa. Breaking spectrum gridlock with cognitive radios: An information theoretic perspective. *Proceedings of the IEEE*, 97(5):894–914, 2009.
79. I. S. Gradshteyn and I. M. Ryzhik. *Table of Integrals, Series and Products*. CA: Academic, San Diego, 6th edition, 2000.
80. G. Gui, H. Huang, Y. Song, and H. Sari. Deep learning for an effective nonorthogonal multiple access scheme. *IEEE Transactions on Vehicular Technology*, 67(9):8440–8450, 2018.
81. Y. Gwon, H. T. Kung, and D. Vlah. Compressive sensing with optimal sparsifying basis and applications in spectrum sensing. In *Global Communications Conference (GLOBECOM)*, pages 5386–5391, Dec 2012.
82. K. Hamdi, X. N. Zeng, A. Ghrayeb, and K. B. Letaief. Impact of noise power uncertainty on cooperative spectrum sensing in cognitive radio systems. In *2010 IEEE Global Telecommunications Conference (GLOBECOM)*, pages 1–5, Dec 2010.
83. Haozhou Xue and Feifei Gao. A machine learning based spectrum-sensing algorithm using sample covariance matrix. In *2015 10th International Conference on Communications and Networking in China (ChinaCom)*, pages 476–480, 2015.
84. G. Hattab and M. Ibnkahla. Multiband spectrum access: Great promises for future cognitive radio networks. *Proceedings of the IEEE*, 102(3):282–306, March 2014.
85. V. Havary-Nassab, S. Hassan, and S. Valaee. Compressive detection for wideband spectrum sensing. In *2010 IEEE International Conference on Acoustics, Speech and Signal Processing*, pages 3094–3097, March 2010.
86. V. Havary-Nassab, S. Hassan, and S. Valaee. Compressive detection for wideband spectrum sensing. In *IEEE International Conference on Acoustics Speech and Signal Processing (ICASSP)*, pages 3094–3097, March 2010.

## BIBLIOGRAPHY

87. S. Haykin, D. J. Thomson, and J. H. Reed. Spectrum sensing for cognitive radio. *Proceedings of the IEEE*, 97(5):849–877, May 2009.
88. Simon Haykin. Cognitive radio: brain-empowered wireless communications. *IEEE Journal on Selected Areas in Communications*, 23(2):201–220, Feb 2005.
89. H. He and H. Jiang. Deep learning based energy efficiency optimization for distributed cooperative spectrum sensing. *IEEE Wireless Communications*, 26(3):32–39, 2019.
90. S. P. Herath, N. Rajatheva, and C. Tellambura. Energy detection of unknown signals in fading and diversity reception. *IEEE Transactions on Communications*, 59(9):2443–2453, September 2011.
91. H. Huang, S. Guo, G. Gui, Z. Yang, J. Zhang, H. Sari, and F. Adachi. Deep learning for physical-layer 5G wireless techniques: Opportunities, challenges and solutions. *IEEE Wireless Communications*, 27(1):214–222, 2020.
92. H. Huang, Y. Song, J. Yang, G. Gui, and F. Adachi. Deep-learning-based millimeter-wave massive MIMO for hybrid precoding. *IEEE Transactions on Vehicular Technology*, 68(3):3027–3032, 2019.
93. H. Huang, J. Yang, H. Huang, Y. Song, and G. Gui. Deep learning for super-resolution channel estimation and DOA estimation based massive MIMO system. *IEEE Transactions on Vehicular Technology*, 67(9):8549–8560, 2018.
94. S. Ji, W. Xu, M. Yang, and K. Yu. 3D convolutional neural networks for human action recognition. *IEEE Transactions on Pattern Analysis and Machine Intelligence*, 35(1):221–231, 2013.
95. M. Jin, Q. Guo, J. Tong, J. Xi, and Y. Li. Energy detection of DVB-T signals against noise uncertainty. *IEEE Communications Letters*, 18(10):1831–1834, 2014.
96. W. Jouini. Energy detection limits under log-normal approximated noise uncertainty. *IEEE Signal Processing Letters*, 18(7):423–426, 2011.
97. S. S. Kalamkar and A. Banerjee. On the performance of generalized energy detector under noise uncertainty in cognitive radio. In *National Conference on Communications (NCC)*, pages 1–5, Feb 2013.
98. S. S. Kalamkar, A. Banerjee, and A. K. Gupta. SNR wall for generalized energy detection under noise uncertainty in cognitive radio. In *2013 19th Asia-Pacific Conference on Communications (APCC)*, pages 375–380, Aug 2013.
99. S. Kapoor, S. Rao, and G. Singh. Opportunistic spectrum sensing by employing matched filter in cognitive radio network. In *2011 International Conference on Communication Systems and Network Technologies*, pages 580–583, 2011.
100. A. Karpathy and L. Fei-Fei. Deep visual-semantic alignments for generating image descriptions. In *2015 IEEE Conference on Computer Vision and Pattern Recognition (CVPR)*, pages 3128–3137, 2015.
101. S.M. Kay. *Fundamentals of Statistical Signal Processing: Detection theory*. Prentice Hall Signal Processing Series. Prentice-Hall PTR, 1998.
102. Steven Kay. *Intuitive Probability and Random Processes using MATLAB*. Springer, 2006.
103. K. Kim, I. A. Akbar, K. K. Bae, J. S. Um, C. M. Spooner, and J. H. Reed.

Cyclostationary approaches to signal detection and classification in cognitive radio. In *2007 2nd IEEE International Symposium on New Frontiers in Dynamic Spectrum Access Networks*, pages 212–215, April 2007.

104. M. Kim and J. i. Takada. Efficient multi-channel wideband spectrum sensing technique using filter bank. In *2009 IEEE 20th International Symposium on Personal, Indoor and Mobile Radio Communications*, pages 1014–1018, Sept 2009.

105. V.I. Kostylev. Energy detection of a signal with random amplitude. In *IEEE International Conference on Communications (ICC 2002)*, volume 3, pages 1606–1610 vol.3, 2002.

106. S. Kusaladharma and C. Tellambura. Aggregate interference analysis for underlay cognitive radio networks. *IEEE Wireless Communications Letters*, 1(6):641–644, 2012.

107. S. Kusaladharma and C. Tellambura. On approximating the cognitive radio aggregate interference. *IEEE Wireless Communications Letters*, 2(1):58–61, 2013.

108. E. G. Larsson and M. Skoglund. Cognitive radio in a frequency-planned environment: some basic limits. *IEEE Transactions on Wireless Communications*, 7(12):4800–4806, 2008.

109. W. Lee, M. Kim, and D. Cho. Deep cooperative sensing: Cooperative spectrum sensing based on convolutional neural networks. *IEEE Transactions on Vehicular Technology*, 68(3):3005–3009, 2019.

110. W.C.Y. Lee. *Mobile Communications Engineering: Theory and Applications*. McGraw-Hill Professional Publishing, 1998.

111. W. M. Lees, A. Wunderlich, P. J. Jeavons, P. D. Hale, and M. R. Souryal. Deep learning classification of 3.5-GHz band spectrograms with applications to spectrum sensing. *IEEE Transactions on Cognitive Communications and Networking*, 5(2):224–236, 2019.

112. Alberto Leon-Garcia. *Probability, Statistics, and Random Processes for Electrical Engineering, Pearson*. Pearson, 2007.

113. J. Li, F. Gao, T. Jiang, and W. Chen. A new spectrum sensing strategy when primary user has multiple power levels. In *2014 IEEE Global Communications Conference*, pages 840–845, Dec 2014.

114. Ye Li and Zhi Ding. Arma system identification based on second-order cyclostationarity. *IEEE Transactions on Signal Processing*, 42(12):3483–3494, Dec 1994.

115. Z. Li, F. R. Yu, and M. Huang. A cooperative spectrum sensing consensus scheme in cognitive radios. In *IEEE INFOCOM 2009*, pages 2546–2550, April 2009.

116. M. Lin and A. P. Vinod. Progressive decimation filter banks for variable resolution spectrum sensing in cognitive radios. In *2010 17th International Conference on Telecommunications*, pages 857–863, April 2010.

117. C. Liu, X. Liu, and Y. Liang. Deep CNN for spectrum sensing in cognitive radio. In *ICC 2019 - 2019 IEEE International Conference on Communications*

## BIBLIOGRAPHY

*(ICC)*, pages 1–6, 2019.
118. C. Liu, J. Wang, X. Liu, and Y. Liang. Deep CM-CNN for spectrum sensing in cognitive radio. *IEEE Journal on Selected Areas in Communications*, 37(10):2306–2321, 2019.
119. Y. Lu, P. Zhu, D. Wang, and M. Fattouche. Machine learning techniques with probability vector for cooperative spectrum sensing in cognitive radio networks. In *2016 IEEE Wireless Communications and Networking Conference*, pages 1–6, 2016.
120. J. Lunden, V. Koivunen, A. Huttunen, and H. V. Poor. Spectrum sensing in cognitive radios based on multiple cyclic frequencies. In *2007 2nd International Conference on Cognitive Radio Oriented Wireless Networks and Communications*, pages 37–43, Aug 2007.
121. J. Lunden, V. Koivunen, A. Huttunen, and H. V. Poor. Collaborative cyclostationary spectrum sensing for cognitive radio systems. *IEEE Transactions on Signal Processing*, 57(11):4182–4195, 2009.
122. A. Lpez-Parrado and J. Velasco-Medina. Cooperative wideband spectrum sensing based on sub-Nyquist sparse fast Fourier transform. *IEEE Transactions on Circuits and Systems II: Express Briefs*, 63(1):39–43, Jan 2016.
123. K. M. Captain and M. V. Joshi. SNR wall under different scenarios for cooperative spectrum sensing using ged. In *2019 11th International Conference on Communication Systems Networks (COMSNETS)*, pages 259–265, 2019.
124. J. Ma, G. Y. Li, and B. H. Juang. Signal processing in cognitive radio. *Proceedings of the IEEE*, 97(5):805–823, May 2009.
125. J. Ma and Y. G. Li. Soft combination and detection for cooperative spectrum sensing in cognitive radio networks. In *IEEE GLOBECOM 2007 - IEEE Global Telecommunications Conference*, pages 3139–3143, Nov 2007.
126. J. Ma, G. Zhao, and Y. Li. Soft combination and detection for cooperative spectrum sensing in cognitive radio networks. *IEEE Transactions on Wireless Communications*, 7(11):4502–4507, November 2008.
127. Y. Ma, Y. Gao, Y. Liang, and S. Cui. Reliable and efficient sub-Nyquist wideband spectrum sensing in cooperative cognitive radio networks. *IEEE Journal on Selected Areas in Communications*, 34(10):2750–2762, 2016.
128. Y. Ma, Y. Gao, Y. C. Liang, and S. Cui. Efficient blind cooperative wideband spectrum sensing based on joint sparsity. In *2016 IEEE Global Communications Conference (GLOBECOM)*, pages 1–6, Dec 2016.
129. K. Maeda, A. Benjebbour, T. Asai, T. Furuno, and T. Ohya. Recognition among OFDM-Based systems utilizing cyclostationarity-inducing transmission. In *2007 2nd IEEE International Symposium on New Frontiers in Dynamic Spectrum Access Networks*, pages 516–523, April 2007.
130. A. Mariani, A. Giorgetti, and M. Chiani. SNR wall for energy detection with noise power estimation. In *2011 IEEE International Conference on Communications (ICC)*, pages 1–6, June 2011.
131. M. T. Masonta, M. Mzyece, and N. Ntlatlapa. Spectrum decision in cognitive radio networks: A survey. *IEEE Communications Surveys Tutorials*,

15(3):1088–1107, 2013.
132. O. Mehanna and N. D. Sidiropoulos. Frugal sensing: Wideband power spectrum sensing from few bits. *IEEE Transactions on Signal Processing*, 61(10):2693–2703, May 2013.
133. M. Mishali and Y. C. Eldar. Blind multiband signal reconstruction: Compressed sensing for analog signals. *IEEE Transactions on Signal Processing*, 57(3):993–1009, March 2009.
134. M. Mishali and Y. C. Eldar. Blind multiband signal reconstruction: Compressed sensing for analog signals. *IEEE Transactions on Signal Processing*, 57(3):993–1009, March 2009.
135. J. Mitola. Cognitive radio for flexible mobile multimedia communications. In *Mobile Multimedia Communications, 1999. (MoMuC '99) 1999 IEEE International Workshop on*, pages 3–10, 1999.
136. J. Mitola. Software radio architecture: a mathematical perspective. *IEEE Journal on Selected Areas in Communications*, 17(4):514–538, Apr 1999.
137. J. Mitola and G. Q. Maguire. Cognitive radio: making software radios more personal. *IEEE Personal Communications*, 6(4):13–18, Aug 1999.
138. F. Moghimi, A. Nasri, and R. Schober. Adaptive $l_p$−norm spectrum sensing for cognitive radio networks. *IEEE Transactions on Communications*, 59(7):1934–1945, July 2011.
139. A. Montazeri, J. Haddadnia, and S. H. Safavi. Fuzzy hypothesis testing for cooperative sequential spectrum sensing under noise uncertainty. *IEEE Communications Letters*, 20(12):2542–2545, 2016.
140. Montgomery and Ruger. *Applied Statistics and Probability for Engineers*. John Wiley, 2006.
141. S. Nallagonda, S. K. Bandari, S. D. Roy, and S. Kundu. Performance of cooperative spectrum sensing with soft data fusion schemes in fading channels. In *2013 Annual IEEE India Conference (INDICON)*, pages 1–6, Dec 2013.
142. S. Nallagonda, A. Chandra, S.D. Roy, S. Kundu, P. Kukolev, and A. Prokes. Detection performance of cooperative spectrum sensing with hard decision fusion in fading channels. *International Journal of Electronics*, 103(2):297–321, 2016.
143. S. Nallagonda, S. D. Roy, A. Chandra, and S. Kundu. Performance of cooperative spectrum sensing in hoyt fading channel under hard decision fusion rules. In *2012 5th International Conference on Computers and Devices for Communication (CODEC)*, pages 1–4, Dec 2012.
144. S. Nallagonda, S. D. Roy, and S. Kumdu. Performance of cooperative spectrum sensing in fading channels. In *2012 1st International Conference on Recent Advances in Information Technology (RAIT)*, pages 202–207, March 2012.
145. E.A. Neasmith and N.C. Beaulieu. New results on selection diversity. *IEEE Transactions on Communications*, 46(5):695–704, May 1998.
146. Albert H. Nuttall. Some integrals involving the $Q_M$- function. Technical report, Naval Underwater Syst. Center (NUSC) Tech. Rep., May 1974.
147. T. J. OShea, T. Roy, and T. C. Clancy. Over-the-air deep learning based radio

signal classification. *IEEE Journal of Selected Topics in Signal Processing*, 12(1):168–179, 2018.
148. H. Palangi, L. Deng, Y. Shen, J. Gao, X. He, J. Chen, X. Song, and R. Ward. Deep sentence embedding using long short-term memory networks: Analysis and application to information retrieval. *IEEE/ACM Transactions on Audio, Speech, and Language Processing*, 24(4):694–707, 2016.
149. A.K. Papazafeiropoulos and S.A. Kotsopoulos. The $\eta$-$\lambda$-$\mu$: A general fading distribution. In *IEEE Global Telecommunications Conference, GLOBECOM 2009*, pages 1–5, Nov 2009.
150. A. Papoulis and S.U. Pillai. *Probability, Random Variables, and Stochastic Processes*. McGraw-Hill series in electrical engineering: Communications and signal processing. Tata McGraw-Hill, 2002.
151. E. C. Y. Peh, Y. C. Liang, Y. L. Guan, and Y. Zeng. Optimization of cooperative sensing in cognitive radio networks: A sensing-throughput tradeoff view. *IEEE Transactions on Vehicular Technology*, 58(9):5294–5299, Nov 2009.
152. Y.L. Polo, Ying Wang, A. Pandharipande, and G. Leus. Compressive wide-band spectrum sensing. In *IEEE International Conference on Acoustics, Speech and Signal Processing (ICASSP)*, pages 2337–2340, April 2009.
153. R. V. Prasad, P. Pawlczak, J. A. Hoffmeyer, and H. S. Berger. Cognitive functionality in next generation wireless networks: standardization efforts. *IEEE Communications Magazine*, 46(4):72–78, April 2008.
154. Z. Quan, S. Cui, H. V. Poor, and A. H. Sayed. Collaborative wideband sensing for cognitive radios. *IEEE Signal Processing Magazine*, 25(6):60–73, November 2008.
155. Z. Quan, S. Cui, A. H. Sayed, and H. V. Poor. Optimal multiband joint detection for spectrum sensing in cognitive radio networks. *IEEE Transactions on Signal Processing*, 57(3):1128–1140, March 2009.
156. Vijay K. Rohatgi and A.K. Md. Ehsanes Salehi. *An Introduction to Probability and Statistics*. Wiley, 2011.
157. Sheldon M. Ross. *Introduction to Probability Models*. Academic Press, 2014.
158. H. Sadeghi and P. Azmi. Cyclostationarity-based cooperative spectrum sensing for cognitive radio networks. In *International Symposium on Telecommunications (IST)*, pages 429–434, Aug 2008.
159. T. N. Sainath, R. J. Weiss, K. W. Wilson, B. Li, A. Narayanan, E. Variani, M. Bacchiani, I. Shafran, A. Senior, K. Chin, A. Misra, and C. Kim. Multichannel signal processing with deep neural networks for automatic speech recognition. *IEEE/ACM Transactions on Audio, Speech, and Language Processing*, 25(5):965–979, 2017.
160. A. Salarvan and G. K. Kurt. Multi-antenna spectrum sensing for cognitive radio under Rayleigh channel. In *2012 IEEE Symposium on Computers and Communications (ISCC)*, pages 000780–000784, July 2012.
161. N. Samuel, T. Diskin, and A. Wiesel. Deep MIMO detection. In *2017 IEEE 18th International Workshop on Signal Processing Advances in Wireless Communications (SPAWC)*, pages 1–5, 2017.

162. R. Sarikaya, G. E. Hinton, and A. Deoras. Application of deep belief networks for natural language understanding. *IEEE/ACM Transactions on Audio, Speech, and Language Processing*, 22(4):778–784, 2014.
163. R. Sarikhani and F. Keynia. Cooperative spectrum sensing meets machine learning: Deep reinforcement learning approach. *IEEE Communications Letters*, 24(7):1459–1462, 2020.
164. M. Schwartz, W.R. Bennett, and S. Stein. *Communication Systems and Techniques*. Wiley, 1995.
165. S. Sedighi, A. Taherpour, S. Gazor, and T. Khattab. Eigenvalue-based multiple antenna spectrum sensing: Higher order moments. *IEEE Transactions on Wireless Communications*, 16(2):1168–1184, 2017.
166. E. Shelhamer, J. Long, and T. Darrell. Fully convolutional networks for semantic segmentation. *IEEE Transactions on Pattern Analysis and Machine Intelligence*, 39(4):640–651, 2017.
167. H. Shin, H. R. Roth, M. Gao, L. Lu, Z. Xu, I. Nogues, J. Yao, D. Mollura, and R. M. Summers. Deep convolutional neural networks for computer-aided detection: CNN architectures, dataset characteristics and transfer learning. *IEEE Transactions on Medical Imaging*, 35(5):1285–1298, 2016.
168. A. Singh, M. R. Bhatnagar, and R. K. Mallik. Optimization of cooperative spectrum sensing with an improved energy detector over imperfect reporting channels. In *2011 IEEE Vehicular Technology Conference (VTC Fall)*, pages 1–5, Sept 2011.
169. A. Singh, M. R. Bhatnagar, and R. K. Mallik. Cooperative spectrum sensing in multiple antenna based cognitive radio network using an improved energy detector. *IEEE Communications Letters*, 16(1):64–67, January 2012.
170. P.C. Sofotasios, E. Rebeiz, Li Zhang, T.A. Tsiftsis, D. Cabric, and S. Freear. Energy detection based spectrum sensing over $\kappa$-$\mu$ and $\kappa$-$\mu$ extreme fading channels. *Vehicular Technology, IEEE Transactions on*, 62(3):1031–1040, March 2013.
171. J. Song, Z. Feng, P. Zhang, and Z. Liu. Spectrum sensing in cognitive radios based on enhanced energy detector. *IET Communications*, 6(8):805–809, May 2012.
172. Y. Song and J. Xie. Proactive spectrum handoff in cognitive radio ad hoc networks based on common hopping coordination. In *2010 INFOCOM IEEE Conference on Computer Communications Workshops*, pages 1–2, 2010.
173. A. Sonnenschein and P. M. Fishman. Radiometric detection of spread-spectrum signals in noise of uncertain power. *IEEE Transactions on Aerospace and Electronic Systems*, 28(3):654–660, Jul 1992.
174. S. Srinivasa and S. A. Jafar. The throughput potential of cognitive radio: A theoretical perspective. In *2006 Fortieth Asilomar Conference on Signals, Systems and Computers*, pages 221–225, 2006.
175. S. Srinivasa and S. A. Jafar. Cognitive radios for dynamic spectrum access - the throughput potential of cognitive radio: A theoretical perspective. *IEEE Communications Magazine*, 45(5):73–79, 2007.

176. S. Srinu and S. L. Sabat. Cooperative wideband spectrum sensing in suspicious cognitive radio network. *IET Wireless Sensor Systems*, 3(2):153–161, June 2013.
177. C. R. Stevenson, G. Chouinard, Z. Lei, W. Hu, S. J. Shellhammer, and W. Caldwell. IEEE 802.22: The first cognitive radio wireless regional area network standard. *IEEE Communications Magazine*, 47(1):130–138, January 2009.
178. G.L. Stüber. *Principles of Mobile Communication*. Kluwer Academic, Norwell, 2nd edition, 2001.
179. H. Sun, W. Y. Chiu, J. Jiang, A. Nallanathan, and H. V. Poor. Wideband spectrum sensing with sub-Nyquist sampling in cognitive radios. *IEEE Transactions on Signal Processing*, 60(11):6068–6073, Nov 2012.
180. H. Sun, A. Nallanathan, S. Cui, and C. X. Wang. Cooperative wideband spectrum sensing over fading channels. *IEEE Transactions on Vehicular Technology*, 65(3):1382–1394, March 2016.
181. H. Sun, A. Nallanathan, S. Cui, and Cheng-Xiang Wang. Cooperative wideband spectrum sensing over fading channels. *IEEE Transactions on Vehicular Technology*, PP(99):1–1, 2015.
182. H. Sun, A. Nallanathan, C. X. Wang, and Y. Chen. Wideband spectrum sensing for cognitive radio networks: a survey. *IEEE Wireless Communications*, 20(2):74–81, April 2013.
183. Hongjian Sun, Wei-Yu Chiu, Jing Jiang, A. Nallanathan, and H.V. Poor. Wideband spectrum sensing with sub-Nyquist sampling in cognitive radios. *IEEE Transactions on Signal Processing*, 60(11):6068–6073, Nov 2012.
184. Z. Sun and J. N. Laneman. Sampling schemes and detection algorithms for wideband spectrum sensing. In *2014 IEEE International Symposium on Dynamic Spectrum Access Networks (DYSPAN)*, pages 541–552, April 2014.
185. Zhanwei Sun and J.N. Laneman. Performance metrics, sampling schemes, and detection algorithms for wideband spectrum sensing. *IEEE Transactions on Signal Processing*, 62(19):5107–5118, Oct 2014.
186. P. D. Sutton, J. Lotze, K. E. Nolan, and L. E. Doyle. Cyclostationary signature detection in multipath Rayleigh fading environments. In *2007 2nd International Conference on Cognitive Radio Oriented Wireless Networks and Communications*, pages 408–413, Aug 2007.
187. P. D. Sutton, K. E. Nolan, and L. E. Doyle. Cyclostationary signatures for rendezvous in OFDM-Based dynamic spectrum access networks. In *2007 2nd IEEE International Symposium on New Frontiers in Dynamic Spectrum Access Networks*, pages 220–231, April 2007.
188. A. Taherpour, M. Nasiri-Kenari, and S. Gazor. Multiple antenna spectrum sensing in cognitive radios. *IEEE Transactions on Wireless Communications*, 9(2):814–823, February 2010.
189. R. Tandra and A. Sahai. SNR walls for signal detection. *IEEE Journal of Selected Topics in Signal Processing*, 2(1):4–17, Feb 2008.
190. T. A. Tang, L. Mhamdi, D. McLernon, S. A. R. Zaidi, and M. Ghogho. Deep learning approach for network intrusion detection in software defined. In

networking *2016 International Conference on Wireless Networks and Mobile Communications (WINCOM)*, pages 258–263, 2016.

191. K. M. Thilina, K. W. Choi, N. Saquib, and E. Hossain. Machine learning techniques for cooperative spectrum sensing in cognitive radio networks. *IEEE Journal on Selected Areas in Communications*, 31(11):2209–2221, 2013.

192. J. Tian, P. Cheng, Z. Chen, M. Li, H. Hu, Y. Li, and B. Vucetic. A machine learning-enabled spectrum sensing method for OFDM systems. *IEEE Transactions on Vehicular Technology*, 68(11):11374–11378, 2019.

193. Z. Tian and G. B. Giannakis. A wavelet approach to wideband spectrum sensing for cognitive radios. In *2006 1st International Conference on Cognitive Radio Oriented Wireless Networks and Communications*, pages 1–5, June 2006.

194. Z. Tian and G. B. Giannakis. Compressed sensing for wideband cognitive radios. In *2007 IEEE International Conference on Acoustics, Speech and Signal Processing - ICASSP '07*, volume 4, pages IV–1357–IV–1360, April 2007.

195. Z. Tian, Y. Tafesse, and B. M. Sadler. Cyclic feature detection with sub-Nyquist sampling for wideband spectrum sensing. *IEEE Journal of Selected Topics in Signal Processing*, 6(1):58–69, Feb 2012.

196. M. A. Torad, A. El-Kassas, and H. El-Hennawy. Cooperative wideband spectrum sensing over Rician and Nakagami fading channels. In *2015 IEEE International Symposium on Signal Processing and Information Technology (IS-SPIT)*, pages 180–185, Dec 2015.

197. J. A. Tropp, J. N. Laska, M. F. Duarte, J. K. Romberg, and R. G. Baraniuk. Beyond Nyquist: Efficient sampling of sparse bandlimited signals. *IEEE Transactions on Information Theory*, 56(1):520–544, Jan 2010.

198. J. Unnikrishnan and V. V. Veeravalli. Cooperative sensing for primary detection in cognitive radio. *IEEE Journal of Selected Topics in Signal Processing*, 2(1):18–27, Feb 2008.

199. Harry Urkowitz. Energy detection of unknown deterministic signals. *Proceedings of the IEEE*, 55(4):523–531, April 1967.

200. R. Venkataramani and Y. Bresler. Perfect reconstruction formulas and bounds on aliasing error in sub-Nyquist nonuniform sampling of multiband signals. *IEEE Transactions on Information Theory*, 46(6):2173–2183, Sep 2000.

201. O. Vinyals, A. Toshev, S. Bengio, and D. Erhan. Show and tell: A neural image caption generator. In *2015 IEEE Conference on Computer Vision and Pattern Recognition (CVPR)*, pages 3156–3164, 2015.

202. E. Visotsky, S. Kuffner, and R. Peterson. On collaborative detection of TV transmissions in support of dynamic spectrum sharing. In *First IEEE International Symposium on New Frontiers in Dynamic Spectrum Access Networks, 2005. DySPAN 2005*, pages 338–345, Nov 2005.

203. B. Wang and K. J. R. Liu. Advances in cognitive radio networks: A survey. *IEEE Journal of Selected Topics in Signal Processing*, 5(1):5–23, 2011.

204. C. Wang and L. Wang. Analysis of reactive spectrum handoff in cognitive radio networks. *IEEE Journal on Selected Areas in Communications*, 30(10):2016–2028, 2012.

205. C. Wang, L. Wang, and F. Adachi. Modeling and analysis for reactive-decision spectrum handoff in cognitive radio networks. In *2010 IEEE Global Telecommunications Conference GLOBECOM 2010*, pages 1–6, 2010.
206. H. Wang, Y. Xu, X. Su, and J. Wang. Cooperative spectrum sensing in cognitive radio under noise uncertainty. In *2010 IEEE 71st Vehicular Technology Conference*, pages 1–5, May 2010.
207. L. Wang and C. Wang. Spectrum handoff for cognitive radio networks: Reactive-sensing or proactive-sensins? In *2008 IEEE International Performance, Computing and Communications Conference*, pages 343–348, 2008.
208. Y. Wang, M. Liu, J. Yang, and G. Gui. Data-driven deep learning for automatic modulation recognition in cognitive radios. *IEEE Transactions on Vehicular Technology*, 68(4):4074–4077, 2019.
209. Y. Wang, A. Narayanan, and D. Wang. On training targets for supervised speech separation. *IEEE/ACM Transactions on Audio, Speech, and Language Processing*, 22(12):1849–1858, 2014.
210. Y. Wang, A. Pandharipande, Y. L. Polo, and G. Leus. Distributed compressive wide-band spectrum sensing. In *Information Theory and Applications Workshop, 2009*, pages 178–183, Feb 2009.
211. S. Wei, D. L. Goeckel, and P. A. Kelly. Convergence of the complex envelope of bandlimited OFDM signals. *IEEE Transactions on Information Theory*, 56(10):4893–4904, Oct 2010.
212. Wei Lin and Qinyu Zhang. A design of energy detector in cognitive radio under noise uncertainty. In *2008 11th IEEE Singapore International Conference on Communication Systems*, pages 213–217, 2008.
213. Inc. Wolfram Research. *Mathematica*. Wolfram Research, Inc., Champaign, Illinois, version 10.1 edition, 2015.
214. Y. Wu, Q. Yang, X. Liu, and K. S. Kwak. Delay-constrained optimal transmission with proactive spectrum handoff in cognitive radio networks. *IEEE Transactions on Communications*, 64(7):2767–2779, 2016.
215. J. Xie, C. Liu, Y. Liang, and J. Fang. Activity pattern aware spectrum sensing: A CNN-based deep learning approach. *IEEE Communications Letters*, 23(6):1025–1028, 2019.
216. T. Xiong, Y. Yao, Y. Ren, and Z. Li. Multiband spectrum sensing in cognitive radio networks with secondary user hardware limitation: Random and adaptive spectrum sensing strategies. *IEEE Transactions on Wireless Communications*, 17(5):3018–3029, 2018.
217. R. T. Yazicigil, T. Haque, M. R. Whalen, J. Yuan, J. Wright, and P. R. Kinget. Wideband rapid interferer detector exploiting compressed sampling with a quadrature analog-to-information converter. *IEEE Journal of Solid-State Circuits*, 50(12):3047–3064, Dec 2015.
218. H. Ye, G. Y. Li, and B. Juang. Power of deep learning for channel estimation and signal detection in OFDM systems. *IEEE Wireless Communications Letters*, 7(1):114–117, 2018.

219. Chia-Pang Yen, Yingming Tsai, and Xiaodong Wang. Wideband spectrum sensing based on sub-Nyquist sampling. *IEEE Transactions on Signal Processing*, 61(12):3028–3040, June 2013.
220. S. K. Yoo, P. C. Sofotasios, S. L. Cotton, S. Muhaidat, O. S. Badarneh, and G. K. Karagiannidis. Entropy and energy detection-based spectrum sensing over $\mathscr{F}$-composite fading channels. *IEEE Transactions on Communications*, 67(7):4641–4653, 2019.
221. T. Young, D. Hazarika, S. Poria, and E. Cambria. Recent trends in deep learning based natural language processing [review article]. *IEEE Computational Intelligence Magazine*, 13(3):55–75, 2018.
222. D. Yu, G. Hinton, N. Morgan, J. Chien, and S. Sagayama. Introduction to the special section on deep learning for speech and language processing. *IEEE Transactions on Audio, Speech, and Language Processing*, 20(1):4–6, 2012.
223. H. Yu, Z. Tan, Y. Zhang, Z. Ma, and J. Guo. DNN filter bank cepstral coefficients for spoofing detection. *IEEE Access*, 5:4779–4787, 2017.
224. T. Yucek and H. Arslan. A survey of spectrum sensing algorithms for cognitive radio applications. *IEEE Communications Surveys Tutorials*, 11(1):116–130, First 2009.
225. F. Zeng, C. Li, and Z. Tian. Distributed compressive spectrum sensing in cooperative multihop cognitive networks. *IEEE Journal of Selected Topics in Signal Processing*, 5(1):37–48, Feb 2011.
226. Jie Zeng and X. Su. On SNR wall phenomenon under cooperative energy detection in spectrum sensing. In *2015 10th International Conference on Communications and Networking in China (ChinaCom)*, pages 53–60, Aug 2015.
227. Y. Zeng, C. L. Koh, and Y. C. Liang. Maximum eigenvalue detection: Theory and application. In *2008 IEEE International Conference on Communications*, pages 4160–4164, May 2008.
228. Y. Zeng and Y. C. Liang. Covariance based signal detections for cognitive radio. In *2007 2nd IEEE International Symposium on New Frontiers in Dynamic Spectrum Access Networks*, pages 202–207, April 2007.
229. Y. Zeng and Y. C. Liang. Maximum-minimum eigenvalue detection for cognitive radio. In *2007 IEEE 18th International Symposium on Personal, Indoor and Mobile Radio Communications*, pages 1–5, Sept 2007.
230. Y. Zeng and Y. C. Liang. Eigenvalue-based spectrum sensing algorithms for cognitive radio. *IEEE Transactions on Communications*, 57(6):1784–1793, June 2009.
231. Y. Zeng and Y. C. Liang. Spectrum-sensing algorithms for cognitive radio based on statistical covariances. *IEEE Transactions on Vehicular Technology*, 58(4):1804–1815, May 2009.
232. Y. Zeng, Y. C. Liang, A. T. Hoang, and E. C. Y. Peh. Reliability of spectrum sensing under noise and interference uncertainty. In *2009 IEEE International Conference on Communications Workshops*, pages 1–5, June 2009.
233. C. Zhang, P. Patras, and H. Haddadi. Deep learning in mobile and wireless networking: A survey. *IEEE Communications Surveys Tutorials*, 21(3):2224–2287, 2019.

234. C. Zhang, C. Yu, and J. H. L. Hansen. An investigation of deep-learning frameworks for speaker verification antispoofing. *IEEE Journal of Selected Topics in Signal Processing*, 11(4):684–694, 2017.
235. K. Zhang, J. Li, and F. Gao. Machine learning techniques for spectrum sensing when primary user has multiple transmit powers. In *2014 IEEE International Conference on Communication Systems*, pages 137–141, 2014.
236. L. Zhang, T. Song, M. Wu, X. Bao, J. Guo, and J. Hu. Traffic-adaptive proactive spectrum handoff strategy for graded secondary users in cognitive radio networks. *Chinese Journal of Electronics*, 24(4):844–851, 2015.
237. W. Zhang and K. B. Letaief. Cooperative spectrum sensing with transmit and relay diversity in cognitive radio networks - [transaction letters]. *IEEE Transactions on Wireless Communications*, 7(12):4761–4766, December 2008.
238. W. Zhang, R. K. Mallik, and K. B. Letaief. Optimization of cooperative spectrum sensing with energy detection in cognitive radio networks. *IEEE Transactions on Wireless Communications*, 8(12):5761–5766, December 2009.
239. X. Zhang, R. Chai, and F. Gao. Matched filter based spectrum sensing and power level detection for cognitive radio network. In *2014 IEEE Global Conference on Signal and Information Processing (GlobalSIP)*, pages 1267–1270, 2014.

# Index

## A
Adaptive spectrum sensing strategy, 59
Additive white Gaussian noise, 44, 56, 67–68, 121, 155
Analog to digital converter, 48, 58
Analog-to-information converter, 60
AND combining, 102, 107–108
Approximation, 129–133
Area under the receiver operating characteristic, 57
Autocorrelation, 27
Autocorrelation function, 28–29
    properties, 29
Autocovariance, 28
Average probability of detection, 70–77
    cooperative spectrum sensing, 74–75
    fading and shadowing, 76–77
    no diversity, 70–72
    square law selection diversity, 72–73

## B
Bandpass filter, 44–47, 68, 85
Bernoulli distribution, 14
Binomial distribution, 14–15

## C
Central limit theorem, 24
Certain event, 2
Channel-by-channel detection, 121
Channel-by-channel square law combining, 125
Channel state information, 124
Characteristic function, 13
Chebyschev inequality, 22
Chi-square distribution, 17
Cognitive radio, 38–43
    interweave, 40
    overlay, 41
    underlay, 40
Complementary cumulative distribution function, 165, 167
Conditional probability, 6
Convolution neural network, 63
Convolutional long short-term deep neural network, 63
Cooperating secondary user, 49, 57–58, 74–75, 80–84, 153, 179, 181
Cooperative spectrum sensing, 49–50, 60–62, 74–75
Cooperative wideband spectrum sensing, 62, 152–153, 179–180
Covariance-based detection, 46
Cumulative distribution function, 8
    joint, 9
    properties, 8
Cyclic autocorrelation function, 45, 55
Cyclostationary detection, 45

## D
Deep cooperative sensing, 63
Deep Reinforcement Learning, 63
Degree of freedom, 123, 159
DeMorgan's rule, 4
Digital video broadcasting-terrestrial, 58
Discrete Fourier transform, 125, 155
Dynamic spectrum management framework, 41–42

## E
Effective rate, 57
Eigenvalue-based detection, 46
Energy detection, 46–47, 49, 52, 55, 61–62, 67–70, 126–128
Ensemble, 25
Equal gain combining, 61, 70, 102, 124
Event(s), 2–3
    independence, 7
Expectation maximization, 62, 153
Expected value, 11–12
Exponential distribution, 16–17

**F**
Fading model, 70
Fast Fourier transform, 62, 125, 155
Federal communications commission, 37
Fusion center, 49, 60, 74, 81, 98, 153, 157–158, 166, 180–181
Fuzzy hypotheses testing, 61

**G**
Gamma distribution, 21–22
Gaussian distribution, 15–16
Gaussian mixture model, 62
Gaussian random process, 34–35
Generalized energy detection, 58, 83, 85
Generalized likelihood ratio test, 55

**H**
Hard combining, 49
Hidden Markov model, 55

**I**
Impossible event, 2
Independent and identically distributed, 23–24, 85, 130, 160
International Telecommunication Union, 37

**J**
Joint detection and estimation, 62, 153
Joint probability density function, 91

**K**
$k$ out of $M$ combining rule, 61, 70, 101
K-nearest-neighbor, 63

**L**
Law of large numbers, 22
    weak, 24
    strong, 24
Linear time invariant, 30–31
Low noise amplifier, 90

**M**
Machine-learning-based spectrum sensing, 62–64
Markov inequality, 22
Matched filter detection, 44–45
Maximal ratio combining, 56, 61
Mean, 11
Mean-square estimation error, 124
Media access control, 42
Modulated wideband converter, 60
Moment generating function, 56
Moments, 11–13
Monte Carlo, 52, 108–112, 122, 154, 168, 174
Multiple primary transmit power, 63, 180

**N**
Naive Bayes classifier, 63
Nakagami distribution, 19–20
Noise uncertainty, 85, 114
Non-central chi-square distribution, 17–18
Null event, 2

**O**
OR combining, 74–75, 99, 102, 107–108
Orthogonal frequency division multiplexing, 58
Outcome, 2

**P**
Partial band Nyquist sampling, 59, 121, 153
pLC diversity, 89, 105
pLS diversity, 96, 106
Power spectral density, 32–34
    output spectral density of an LTI systems, 34
    properties, 33
Primary user, 38, 40–47
Probability, 2–7
    axiomatic definition, 4
    classical definition, 5
    conditional, 6
    relative frequency definition, 5
Probability density function 9
    conditional, 10

INDEX

   joint, 9
   properties, 9
Probability mass function, 122
Probability of detection, 44, 69–70
Probability of false alarm, 44, 69–70

**Q**
Quality of service, 41

**R**
Radio frequency, 37
Random process, 25–31
   definition, 26
   statistics, 26
   strict-sense stationary, 28
   through linear system, 30–31
   wide sense stationary, 28
Random spectrum sensing strategy, 59
Random variable, 7–9
   continuous, 15
   discrete, 14
Ranked channel detection, 121, 154, 156, 162, 179
Ranked square law combining, 126, 135–138
Ranked square law selection, 127, 138–139
Rayleigh distribution, 19
Receiver operating characteristic. 57, 77

**S**
Sample mean, 24
Sample space, 2
Secondary user, 37, 38
SNR wall, 57–58, 83–84, 86
Soft combining, 49, 52, 102, 108, 157
Sparse Bayesian learning, 60

Spectrum analysis, 42
Spectrum correlation function, 55
Spectrum decision, 42
Spectrum mobility, 42
Spectrum policy task force, 37
Spectrum sensing, 42–51
   cooperative, 49–50
   machine-learning-based, 51
   narrowband, 43–47
   wideband, 47–49
Spectrum sensing capability, 59
Square law combining, 57, 61, 70, 84, 124–125, 156–157
Square law selection diversity, 72–73
Standard deviation, 13
Statistical independence, 11
Stochastic process, 25–31
   definition, 25
   statistics, 25
   strict-sense stationary, 28
   wide sense stationary, 28
Support vector machine, 63

**U**
Ultra-high frequency, 47
Uniform distribution, 20–21

**V**
Variance, 13

**W**
White noise, 35–36
   Gaussian noise, 35
Wideband spectrum sensing, 47–49, 58–60
   Nyquist, 48
   sub-Nyquist, 48–49

Printed in the United States
by Baker & Taylor Publisher Services

Printed in the United States
by Baker & Taylor Publisher Services